MW00973309

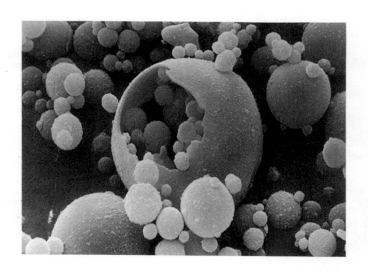

Fly Ash in Concrete

Advances in Concrete Technology

A series edited by V. M. Malhotra
Advanced Concrete Technology Program, CANMET
Ottawa, Ontario, Canada

This series will consist of approximately ten short, sharply focused tracts, each one covering one of the many aspects of concrete technology: materials, construction, and testing. The goal of this series is to provide a convenient, practical, and current source on concrete technology for practicing civil and structural engineers, concrete technologists, manufacturers, suppliers, and contractors involved in construction and maintenance of concrete structures.

This book is part of a series. The publisher will accept continuation orders which may be cancelled at any time and which provide for automatic billing and shipping of each title in the series upon publication. Please write for details.

Fly Ash in Concrete
Production, Properties and Uses

R. C. Joshi

and

R. P. Lohtia

The University of Calgary
Alberta, Canada

GORDON AND BREACH SCIENCE PUBLISHERS

Australia Canada China France Germany India Japan
Luxembourg Malaysia The Netherlands Russia Singapore
Switzerland Thailand United Kingdom

Amsteldijk 166
1st Floor
1079 LH Amsterdam
The Netherlands

British Library Cataloguing in Publication Data

Joshi, Ramesh C.
  Fly ash in concrete : production, properties and uses.–
  (Advances in concrete technology ; v. 2)
  1.Fly ash   2.Concrete
  I.Title   II.Lohtia, Rajinder P.
  666.8'93

ISBN   90-5699-580-4

CONTENTS

3 USE OF FLY ASH IN CEMENT AND CONCRETE

4 EFFECTS OF FLY ASH ON THE PROPERTIES OF FRESH CONCRETE

11 SPECIAL PROBLEMS INCLUDING USE CONSTRAINTS

PREFACE

The industrial revolution increased the use of coal as a fuel. Initially, coal in lumps of 70 to 100 mm was fed into stationary or travelling-grate furnaces. The majority of the residue was collected from the furnace bottom and contained large amounts of unburned coal. About 10% of the ash escaped with the flue gases and caused extensive air pollution.

In late 1920s and early 1930s, pulverized coal was fed into cylindrical vertical furnaces. Coal combustion residue in such furnaces was much finer. Bottom ash accounted for less than 20 to 30% of total ash, since fine ash particles were carried with flue gases through smokestacks into the atmosphere. In Europe and the United Kingdom the ash was always referred to as pulverized fuel ash. But in the United States the ash was termed fly ash because it escaped with the flue gases and "flew" into the atmosphere.

Several methods were employed to separate almost all the fly ash from the flue gases. Collected bottom ash as well as fly ash were invariably sluiced into storage ponds. As land costs soared and water and air pollution problems were recognized, better storage and transportation methods had to be devised.

Engineers, in general, do not like to waste even a by-product if it can be used. Ash was used as a fill and pozzolan even before scientific studies proved that the fly ash portion of the ash is a true pozzolan and can be used to produce economical and durable concrete. In the published literature, the credit goes to R. E. Davis (1937) and his team at Berkeley for suggesting the similarity between volcanic ash and fly ash and their effects when used as a pozzolan or additive to cement concrete.

It has been an uphill battle for fly-ash marketers and others to find increased uses for the ash. Many universities, the U.S. Bureau of Mines, and other government agencies have participated in finding new uses for coal ash. Iowa State University, my alma mater, was one of the pioneers in fundamental and applied research on fly ash, not only for use in concrete but also in soil stabilization and sand–fly ash brick production.

During the last three decades, the use and production of fly ash has increased manyfold. Many new uses have been developed for fly ash in producing new materials; however, the largest single use of fly ash to economize construction costs is in cement concrete. Unless all the ash can be converted into a cementitious material, transportation costs involved in different uses will not allow its increased use in construction.

This book presents a brief history of the uses of fly ash and variations in the properties of ash. If the information provided here can be of assistance to anyone in developing new uses or enhancing current uses, its purpose will have been well served.

A bibliography is included at the end of the book, but there are many other references which have not been included. CANMET, under the direction of V. M. Malhotra, has contributed significantly in disseminating the information on fly ash in concrete.

Various chapters have been arranged to allow the reader to refer to the material of significance. A book like this can never be perfect, but an attempt has been made to assemble information of use to the fly ash user as well as the researcher.

R. C. Joshi

ACKNOWLEDGMENTS

V. M. Malhotra deserves credit for asking me to compile a book on fly ash. Both he and the publisher persevered through many delays.

Jane Wang, Jieren Li, and Amanda Morrall were responsible for typing and artwork for the book, and their assistance is greatly appreciated. The artwork could not have been done without Jane Wang. Special thanks go to Jieren Li, who assisted in the final compilation, typing, and artwork. Thanks also to Chang Qing (Max) Wang for his help in reading the page proofs.

My graduate students, who toiled and produced new data on fly ash over the years, deserve mention. Their efforts allowed me to pursue fundamental and applied research in fly ash.

R. C. Joshi

CHAPTER 1

Introduction

1.1. GENERAL

Fly ash, like volcanic ash, is produced in, manmade small scale volcano like, furnaces of coal burning power plants. Volcanic ashes and other similar natural and manmade materials have been used to produce cementing materials by mixing with lime since ancient times. Romans, Chinese and Indians have used volcanic ashes and pulverised burned bricks for producing highly cementitious mortars in construction of ancient monuments still in existence.

Fly ash is made up of very fine, predominantly spherical glassy particles collected in the dust collection systems from the exhaust gases of fossil fuel power plants. It is generally finer than Portland cement. The major chemical constituents in fly ash are silica, alumina and oxides of iron and calcium. Because of its fineness and pozzolanic and sometimes self cementitious nature, fly ash is widely accepted and specified as mineral admixture in cement and concrete. Fly ash has also been successfully used in many other applications in civil engineering construction and other specialty materials.

Coal is used as fuel for about 40 to 50% of electric power generation all over the world. Figure 1.1 shows a schematic diagram of fly ash production and collection in modern thermal power plant. During the combustion of pulverized coal in suspension-fired furnaces of modern thermal power plants, the volatile matter is vaporized and the majority of the carbon is

1

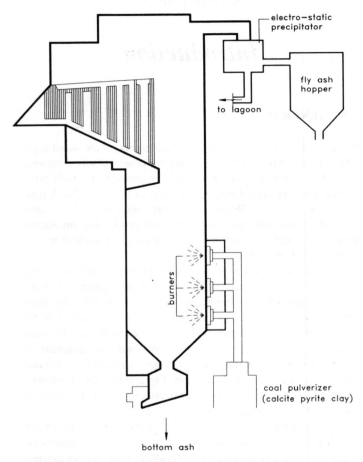

FIGURE 1.1   Schematic diagram showing fossil fuel furnace and fly
ash collection system.

burned off. The mineral matter associated with the coal, such as
clay, quartz, and feldspar disintegrate or slag to varying degree.
The slagged particles and unburned carbon are collected as ash.

The coarser particles fall in the bottom of the furnace and are collected as bottom ash or boiler slag. The finer particles that escape with flue gases are collected as fly ash using cyclone separators, electrostatic precipitators or bag houses. The term economizer ash is applied to the ash collected from the economizer units in many of the present day power plants. In the last three decades or so flue gas desulphurization equipment has been mandated to capture SO_x from the flue gases. The flue gases along with the fly ash are treated using lime or limestone powder or slurry. The majority of the fly ash in such power plants is collected with the calcium sulphate or sulphite sludge or calcium sulphate powder.

Depending upon the collection system, varying from mechanical to electrical precipitators or bag houses and fabric filters, about 85 to 99.9% of the ash from the flue gases is retrieved in the form of fly ash. Fly ash, accounts for 75 to 85% of the total coal ash, and the remainder is collected as bottom ash or boiler slag. Fly ash because of its mineralogical composition, fine particle size and amorphous character is generally pozzolanic and in some cases also self cementitious. The bottom ash and boiler slag are much coarser and are not pozzolanic in nature. It is thus important to recognize that all the ash is not fly ash and the fly ashes produced by different power plants are not equally pozzolanic and, therefore, are not always suitable for use as mineral admixture in concrete.

Fly ash generated in coal burning power plants is an inherently variable material because of several factors. Among these are the type and mineralogical composition of the coal, degree of coal pulverization, type of furnace and oxidation conditions including air-to-fuel ratio, and the manner in which fly ash is collected, handled and stored before use. Since no two utilities or plants may have all of these factors in common, fly ash from various power plants is likely to be different. The fly ash properties may also vary within the same plant because of load conditions over a twenty four hour period. Nonuniformity of fly ash is a serious disadvantage and sometimes is the main hurdle in the effective and wide scale utilization of fly ash as a pozzolan or a cementitious component in cement and concrete.

1.2. HISTORICAL BACKGROUND

Fly ash was recognized as pozzolanic ingredient for use in concrete as early as in 1914. However, the earliest comprehensive study on the use of fly ash in concrete was conducted by Davis et al. (1937). Abdun Nur (1961) compiled data on the properties and uses of fly ash from the literature from 1934 to 1959 including an annotated bibliography. Several other extensive review papers on the use of fly ash in concrete have also been published over the years (Synder 1962, Joshi 1979, Berry and Malhotra 1985 and 1987, Swamy 1986).

Much of the literature concerning use of fly ash in concrete before 1980 dealt with fly ashes resulting from the burning of bituminous coal, designated as Class F pozzolans in ASTM specifications. During the early period of 1940 to 1960, usefulness of only Class F fly ash was investigated since that was the generally available ash at that time. Uses of fly ash were established for a number of applications and the advantages and disadvantages were identified. Co-operative tests were conducted by ASTM Committee C-9 (1962) and studies on the fundamental characteristics of Class F fly ashes were reported by Minnick (1959) during this period. The early studies concluded that substantial amount of the Portland cement in concrete could be replaced with fly ash without adversely affecting the long term strength of concrete (Timms and Grieb 1956).

In the early 1960s self hardening property was detected in some of the fly ashes in the midwestern United States. These were basically sub-bituminous ashes. Since 1970s a number of studies have been reported dealing with the characteristics of the fly ashes from sub-bituminous or "Western" coals in the North American continent. These self hardening fly ashes generally contained larger amounts of calcium as compared to the bituminous ashes. Therefore these ashes did not always meet the ASTM specifications developed for the bituminous ashes produced in the industrial heart land of the United States and all along the eastern seaboard. A new class of ash was therefore added to the ASTM specifications which included ashes with more than 15% calcium oxide and combined silica, alumina, and iron oxide content less than 75%. These ashes exhibit many other useful properties.

Renewed interest in the use of fly ash as an ingredient in concrete in the United States was generated by the passage of the Resource Conservation and Recovery Act (RCRA) in 1976 and the subsequent decision by the Environmental Protection Agency (EPA) to establish guidelines for federal procurement of cement and concrete containing fly ash. On the basis of the guideline published in its final form in the federal register of January 28, 1983, the Federal Highway Administration (FHWA) issued a memorandum to all the state highway agencies in January 1983:

> The guideline requires that all affected agencies revise their specifications, standards, and procedures to remove any discrimination against the use of fly ash in cement and concrete unless such use is found to be technically inappropriate in a particular application. A finding of technical inappropriateness should be documented, open for public scrutiny and a review process established to settle any disagreements. (FHWA 1983)

The most use and, therefore, most research has been on the use of fly ash in cement and concrete. The ACI Committee 226 Report (1987) includes a list of references covering almost all aspects of the use of fly ash in concrete and generally represents the state of the art as of 1985. This report covers the use of fly ash in ready mixed concrete, concrete pavements, concrete for pumping, and in mass concrete. Use of fly ash in concrete masonry units, grouts, mortars, and special applications is also covered in the report.

Canadian Centre for Mineral and Energy Technology (CAN-MET) has published a compilation of abstracts of papers from recent international conferences and symposia on fly ash in concrete (1985) and a comprehensive state of the art report incorporating most of the advancements in the use of fly ash made between 1976 and 1983 (CANMET 1986).

The published proceedings of the workshops on research and development needs for use of fly ash in cement and concrete held in 1981 and 1982 and the project reports on classifications of fly ash for use in cement and concrete by the Electric Power Research Institute (EPRI), 1984 and 1987, are also good sources of recent information.

The American Coal Ash Association (ACAA), formerly known as the Fly Ash Association, every year since 1967, with other agencies such as the Department of Energy (DOE) and EPRI (1995), has been sponsoring international symposia on ash use; and the published proceedings of these symposia provide useful information on numerous potential ash utilization options, particularly in cement and concrete products. During the last two decades, CANMET in association with the American Concrete Institute (ACI) and several other organizations have organized five international conferences on a global basis dealing with the use of fly ash, silica fume, slag and natural pozzolans in concrete. The proceedings of these conferences have been edited by Malhotra (1983, 1986, 1989, 1992 and 1995) and published by ACI. These publications contain a vast number of papers on the use of fly ash in concrete. The authors (Joshi and Lohtia 1993b) have also published a summary of recent investigations on properties of fly ash and their effect on concrete properties as chapters in the series edited by Sarkar (1993) and Ramachandran (1995).

Over the years, extensive research has been conducted throughout the world to investigate the effects of fly ash on properties of both fresh and hardened concrete and on mix proportioning. Fundamental research relating to physical, chemical and mineralogical characterizations of different types

of fly ashes has also been conducted for relating the performance of concrete to the characteristics of fly ashes. This book is
aimed primarily to provide information on the broad aspects of
fly ash utilization in cement and concrete. Although related and
of considerable interest to civil engineers, the use of fly ash and
cement combinations in base courses, embankments, dams and
dikes, and so forth is not covered in detail.

1.3. COAL ASH PRODUCTION

Of the various industrial wastes, coal ash is perhaps the most
abundant world over. For 1989, the International Energy Annual
(1991) indicates that the total coal production worldwide was
4636 million tonnes. The largest producers were China, the
USA, the former USSR, East Germany and Poland. According
to Coal Ash Production and Utilization, in the world, approximately 652 million tonnes of coal ash were produced in 1989,
of which only about 90 million tonnes or 13.8% utilized in
concrete and other civil engineering construction. The total
amount of fly ash used in concrete was about 27.9 million
tonnes, including 2.8 million tonnes as raw material for cement
production, 7.6 million tonnes in blended cement, and 17.5
million tonnes for cement replacement. In 1989, the former
USSR had the largest production of coal ash, about 125 million
tonnes.

The power plants using suspension-fired furnaces in the
United States generated 65.2 million tonnes of coal ash in 1989.
But only 16 million tonnes or about 25% of the total ash produced was utilized. However, the annual utilization of 6.1 million tonnes of fly ash in concrete in the United States was the
highest in the world. Most of the fly ash in the United States
has been used as a mineral admixture or for cement replacement
in concrete construction and related products.

The current annual production of coal ash worldwide is estimated around 600 million tonnes, with fly ash constituting
about 500 million tonnes at 75 to 85% of the total ash produced.
The present day utilization of ash on worldwide basis varies

widely from a minimum of 3% to a maximum of 57%, yet the world average only amounts to about 16% of the total ash. A substantial amount of ash is still disposed of in landfills and/or lagoons at a significant cost to the utility companies and thus to the consumers. All those responsible for disposal and/or utilization of ash are constantly seeking potential ash utilization options. With the effective utilization of increased volumes of fly ash as construction material in civil engineering applications, the problems and costs associated with the ash disposal in an environmentally sound manner will be appreciably mitigated or in some cases possibly eliminated. Currently sustained efforts are underway to investigate many promising uses for increased utilization of fly ash in different fields.

As of today, the utilization of fly ash in concrete is the most extensive and widespread throughout the world as compared to other potential uses of ash. The state of the art is well established with respect to fly ash as a raw material for cement production, for use in blended cements, and as a partial replacement for cement in concrete. Better understanding exists of the advantages and disadvantages of the use of fly ash in concrete. Throughout the world, the use of fly ash in cement and concrete is a relatively highly developed end use. Because of quality control and quality assurance procedures now being adopted, better and improved quality fly ashes are being produced. As a result, there is widespread understanding about the characterization, classification and specifications for use of fly ash in concrete.

Power generating companies marketing fly ash, suitable for use as pozzolan in concrete, are now making special efforts and employing special equipment and procedures to produce high quality fly ash to meet the prescribed specifications. The results of 1984 survey of the National Ready Mixed Concrete Association in the United States showed that 24 to 31% of the concrete produced contains fly ash. The survey further showed that 39% of all ready mixed concrete producers use fly ash. And in the companies using the fly ash, 42% of the concrete produced contains fly ash. Generally large concrete producers make more use of fly ash than do small producers. Most producers believe that the use of fly ash in concrete will remain constant or increase in future years.

At present on the North American continent, including the United States and Canada, the annual utilization of fly ash, mostly as replacement of cement in concrete, amounts to about 12–15% of the total ash produced which is quite significant compared to worldwide use of 6 to 10%. Although in Europe, ash utilization for various purposes amounts to 60–90%, the use in cement production and cement replacement in concrete amounts to only about 10%. It is apparent that in coming decades, all those involved in coal ash disposal in an environmentally acceptable manner, directly or indirectly will confront the public outcry to accelerate progress towards the increased levels of ash utilization. Such efforts will be necessary to prevent air, water and soil pollution, and conserve energy as well as other natural resources.

1.4. COAL COMBUSTION BY-PRODUCTS

In order to have a general understanding of the by-products resulting from the combustion of coal in power plants, the following definitions are provided.

1.4.1. Dry Bottom Ash

This is the residue from coal burned in dry bottom furnaces and is the product that falls through open grates. It generally consists of fused ash particles varying in size from 19 mm to 75 microns. Some of the aggregated fused particles can be easily crushed between fingers, others are hard to break and need mechanical equipment to pulverize further. Many of the particles are spongy and porous and are, therefore, susceptible to deterioration under loading and compaction. The specific gravity of bottom ash particles ranges between 2.08 and 2.73. The major chemical compounds present are silica (SiO_2), ferric oxide (Fe_2O_3) and alumina (Al_2O_3) in varying proportions depending upon the source of coal burned. It is not suitable as a pozzolan for use in cement and concrete. Sometimes it may be used as replacement of fine aggregate in concrete products.

1.4.2. Wet Bottom Boiler Slag

The molten residue in a wet-bottom boiler discharged into a water filled hopper is termed Wet Bottom Boiler Slag. The particle size is smaller than that of dry bottom ash and the particles are glossy and very hard and brittle. Its colour is uniformly black and specific gravity lies in the range of 2.60 to 3.85 depending on the iron oxide (Fe_2O_3) content. The chemical composition of the wet bottom boiler slag is generally the same as that of the dry bottom ash, with the amount of ash varying upon the coal source. Like dry bottom ash, it is not pozzolanic and is generally used as replacement of fine aggregate in concrete products or other civil engineering applications.

1.4.3. Economizer Ash

It consists of coarse particles very similar to fly ash collected from the gases escaping the boiler using electrostatic precipitators and hoppers below the economizer unit. It is much finer than bottom ash as well as wet bottom boiler slag. The economizer ash is generally not suitable for use as a pozzolan and needs to be disposed of or utilized as filler in specialty materials. In case the fly ash from a power plant is not utilized as mineral admixture in concrete, economizer ash can be combined with fly ash for disposal or other uses such as filler material.

1.4.4. Fly Ash

This is the material collected from the flue gases using electro-static or mechanical precipitators or bag houses. Fly ash is separated from the exhaust gases of power plants with suspension-fired furnace in which pulverized coal is used as fuel. It is generally finer than Portland cement and consists of small glassy spheres ranging from less than 1 to more than 100 microns. The major components of fly ash reported in oxide form are silica (SiO_2), alumina (Al_2O_3), and oxides of calcium and iron (CaO and Fe_2O_3). Fly ash composition varies with the source of coal. Because of its fineness and mineralogy including amorphous nature, fly ash is generally pozzolanic and sometimes also self cementitious. Fly ash has been established and is widely used as mineral admixture or supplementary cementing material in cement concrete to impart specific properties to concrete for field applications.

1.4.5. Flue Gas Desulphurization or Scrubber Sludge

It is the residue resulting from flue gas desulphurization, sometimes called desulphurization gypsum or simply waste gypsum. Limestone powder or slurry is used to capture SO_x from the flue gases. Fly ash may or may not be separated prior to desulphurization. The characteristics of desulphurization ashes depend on their sulphate, sulphite and lime content. The fixated flue gas desulphurization (FGD) sludge has a great potential for use as construction material in combination with fly ash, lime and cement. Research is underway for developing uses for FGD to reduce surface subsidence due to abandoned deep coal mines and to control acid mine drainage.

1.5. FLY ASH UTILIZATION PRACTICES

The best method of disposing a waste by-product is to use it in one form or the other. Utilization of coal combustion by-products is therefore becoming an increasingly attractive alternative to disposal all over the world for the following reasons.

- Costs and problems associated with the disposal in an environmentally sound manner are minimized or eliminated.

- Less area is required for disposal, thus enabling other uses of the land and decreased permit requirements.

- There may be financial returns from the coal ash sales, or at least processing costs may be offset.

- Use of coal ash can replace some scarce or expensive natural resources.

- Conservation of non-renewable energy source required in processing and transportation of ash for disposal.

Essentially coal ash utilization is classified in the following schemes:

(1) **High Volume/Low Technology**
For example: as structural fills in embankments, dams, dikes and levees, and as sub-base and base courses in road way construction.

(2) **Medium Volume/Medium Technology**
This includes the use of coal ash, particularly fly ash as raw material in cement production, as an admixture in blended cements and as replacement of cement or as a mineral admixture in concrete. In addition coal ash including fly ash may be used as partial replacement of fine aggregate in concrete. Fly ash is also used for producing light weight aggregates for concrete and many other applications.

(3) **Low Volume/High Technology**
This includes the coal ash utilization in high value added applications such as metal extractions. High value metal recovery of Aluminum (Al), Gold (Au), Silver (Ag), Vana-

dium (Va) and Strontium (Sr) fall in this category. Fly ash has potential uses for producing light weight refractory material and exotic high temperature resistant tiles. Cenospheres or floaters in fly ash are used as specialty refractory material as well as additives to forging to produce high strength alloys. Light weight and high strength "Ashalloys" prepared by blending cenospheres with aluminum melt have been developed. These alloys are now commonly used in automobile and aircraft industry.

(4) **Miscellaneous Uses**

Based upon its physical properties, coal ash is used as land fill for land reclamations for residential, commercial and recreational development projects. Coal ash also finds uses as filler in asphalt, plastics, paints and rubber products. Coal ash has been successfully used in water treatment and as absorbent for oil and chemical spills.

Other potential uses of coal ash are in mineral wool, gyprock wall boards and in the manufacture of building components such as concrete masonry blocks, bricks and tiles (Frigione et al. 1993b). Fly ash in combination with lime or cement has been successfully used for solidification and stabilization of oil/gas well and other chemical sludges as well as for leachate migration control from oil waste dumps and sludge ponds. Oil well cement used for cementing the casing pipe and for filling the annular space between the casing and formation strata also utilize fly ash as admixture to reduce the specific gravity of the cement–water grout.

Many of the uses of fly ash are directly related to its pozzolanic properties. It is thus considered a valuable construction material in its own right, not only as a substitute or replacement. Most use, and therefore, most research to date, has been for fly ash in cement concrete. Fly ash was initially used, in 1940s and 1950s, as finely divided pozzolanic material in mass concrete structures, primarily dams, for reducing the heat of hydration. Many buildings and highways were also constructed with fly ash concrete, using fly ash for partial replacement of cement. Fly ash use in cement manufacture and in structural building products continued to increase through the 1970s and 1980s.

Since 1983 when the Resource Conservation and Recovery Act (RCRA), encouraging recovery and reuse of waste products, was passed fly ash utilization has received even more impetus. Worldwide, extensive research programs are underway to foster and promote dissemination of information. Numerous international conferences/symposia on use of coal ash in concrete manifest the increased interest and concern of all those responsible for ash disposal/utilization.

CHAPTER 2

Types and Properties
of Fly Ash

2.1. GENERAL

According to ASTM C618-93 specification (1993) for "Fly Ash and Raw or Calcined Natural Pozzolan for use as Mineral Admixture in Portland Cement Concrete," pozzolans are defined as "silicious and aluminous materials which in themselves possess little or no cementitious value but will, in finely divided form and in the presence of moisture, chemically react with calcium hydroxide at ordinary temperatures to form compounds possessing cementitious properties." Fly ash like volcanic ash, a natural pozzolan, has been established and successfully used as a finely divided pozzolanic material in cement concrete and other related products for more than half a century. When fly ash is used in combination with Portland cement, calcium hydroxide liberated from the hydration of Portland cement reacts with the alumino-silicates present in the fly ash to form cementitious compounds possessing cohesive and adhesive properties. These calcium alumino-silicate hydrates so formed are termed as pozzolanic reaction products. However, pozzolanic reactions are much slower than cement hydration reactions.

ASTM C618-93 categorizes natural pozzolans and fly ashes into the following three categories.

Class N: Raw or calcined natural pozzolans such as some diatomaceous earths, opaline chert and shale, stuffs, volcanic ashes and pumice are included in this category. Calcined kaolin clay and laterite shale also fall in this category of pozzolans.

Class F: Fly ash normally produced from burning anthracite or bituminous coal falls in this category. This class of fly ash exhibits pozzolanic property but rarely, if any, self hardening property.

Class C: Fly ash normally produced from lignite or sub-bituminous coal is the only material included in this category. This class of fly ash has both pozzolanic and varying degree of self cementitious properties. (Most Class C fly ashes contain more than 15% CaO. But some Class C fly ashes may contain as little as 10% CaO).

Table 2.1 presents chemical and physical requirements for fly ash and natural pozzolans for use as a mineral admixture in Portland cement concrete. The table also includes a list of procedures and materials used for assessing the quality of fly ash/natural pozzolans to meet the requirements of ASTM C618.

2.2. CLASSIFICATION

As indicated in Section 2.1, according to ASTM C618, two major classes of fly ash are recognized. These two classes are related to the type of coal burned and are designated Class F and Class C in most of the current literature. Class F fly ash is normally produced by burning anthracite or bituminous coal while Class C fly ash is generally obtained by burning sub-bituminous or lignite coal. The important characteristics of these two types of ashes are discussed below.

Presently, no appreciable amount of anthracite coal is used for power generation. Therefore, essentially all Class F fly ashes

TABLE 2.1   Requirements for fly ash and natural pozzolans for use as mineral admixtures in Portland cement concrete as per ASTM C618-93

Requirements	Mineral admixture class		
	N	F	C
Chemical requirements			
$SiO_2 + Al_2O_3 + Fe_2O_3$, min%	70.0	70.0	50.0
SO_3, max%	4.0	5.0	5.0
Moisture content, max%	3.0	3.0	3.0
Loss on ignition, max%	10.0	6.0	6.0
Physical requirements			
Amount retained when wet sieved on 45 µm sieve, max%	34	34	34
Pozzolanic activity index, with Portland cement at 28 days, min% of control	75	75	75
Pozzolanic activity index, with lime, at 7 days, min (MPa)	5.5	5.5	–
Water requirement, max% of control	115	105	105
Autoclave expansion or contraction, max%	0.8	0.8	0.8
Specific gravity, max variation from average %	5	5	5
Percent retained on 45 µm sieve, max variation, percentage points from average	5	5	5
Optional requirements			
Available alkalies, as Na_2O, max%	1.5	1.5	1.5
Increase of drying shrinkage of mortar bars at 28 days, max%	0.03	0.03	0.03
Reduction of mortar expansion at 14 days in alkali reactivity test, min%	75	–	–
Mortar expansion at 14 days in alkali reactivity test, max%	0.02	0.02	0.02
Air entrainment required for air content of 18.0 vol%, variation max%	20	20	20

presently available are derived from bituminous coal found in the midwest and eastern states of the United States and some of the eastern provinces of Canada. Class F fly ashes with calcium oxide (CaO) content less than 6%, designated as low calcium ashes, are not self hardening but generally exhibit pozzolanic properties. These ashes contain more than 2% unburned carbon determined by loss on ignition (LOI) test. Quartz, mullite and hematite are the major crystalline phases identified in North American fly ashes, derived from eastern bituminous coal. Essentially, all the fly ashes and, therefore, most research concerning use of fly ash in cement and concrete in the United States before 1975 dealt with Class F fly ashes.

In the presence of water, the fly ash particles produced from a bituminous coal react with lime or calcium hydroxide to form cementing compounds similar to those generated on the hydration of Portland cement. Previous research findings and majority of current industry practices indicate that satisfactory and acceptable concrete can be produced with the Class F fly ash replacing 15 to 30% of cement by weight. When Class F fly ash is used for producing air entrained concrete to improve freeze–thaw durability, the demand for air entraining admixtures is generally increased. Use of Class F fly ash in general reduces water demand as well as heat of hydration. The concrete made with Class F fly ash also exhibits improved resistance to sulphate attack and chloride ion ingress.

Class C fly ashes, containing usually more than 15% CaO and also called high calcium ashes, became available for use in concrete industry only in the last 20 years or so with the openings of Western Coal Fields in the 1970s. This class of ash is typically derived from Wyoming and Montana sub-bituminous coal or North Dakota and Texas lignite coal in the United States. Class C fly ashes are not only pozzolanic in nature but are invariably self cementitious. When mixed with water, Class C ashes hydrate almost in the same way as Portland cement does. In many cases this initial hardening occurs relatively fast. The degree of self hardening generally varies with the calcium oxide content of the fly ash. Higher CaO content generally denotes higher self cementitious value for Class C fly ash. Since 1980, a number of studies have reported the characteristics and the influence of Class C fly ash on concrete properties.

Class C ashes have very little unburned carbon with loss on ignition (LOI) being less than 1%. The typical crystalline phases of these ashes are anhydride, tricalcium aluminate, lime, quartz, periclase, mullite, merwinite and ferrite.

Ashes typically derived from lignites, sub-bituminous and bituminous coal from the eastern flats of Rocky Mountain region and midwestern region of the North American Continent, contain calcium oxide content ranging from 7 to 30%. These ashes, unlike the ashes produced from bituminous coal from eastern region of the North American Continent, exhibit varying degree of self hardening in addition to pozzolanic activity. These ashes defy exclusive classification as per the existing ASTM classification guidelines and need to be studied on a case by case basis regarding their use in concrete. The main minerals present in western and midwestern ashes are tricalcium silicate (alite), dicalcium silicate (belite), portlandite, calcite, hematite, sodium sulphate (thenardite), and quartz. During the last 10–15 years, extensive studies as well as field applications have demonstrated that successful and acceptable concrete can be produced using 50 to 60% of these Class C or marginally Class C fly ashes for cement replacement in conjunction with chemical admixtures (Joshi et al. 1986, 1991, 1993).

The ASTM classification of fly ashes based on the type of coal burned does not seem to be realistic and rational in predicting the behaviour of fly ashes when used in concrete (Joshi and Lohtia 1995). Furthermore, in the ASTM C618 specification, the arbitrary chemical requirement for the sum of ($SiO_2 + Al_2 + Fe_2O_3$) does not seem to be rational and convincing, as this has no direct relationship to the properties of the material. Wide differences in characteristics are observed within each class. Class F fly ash can be produced from coals that are not bituminous, for example, some of sub-bituminous coals from Alberta,

TABLE 2.2   Classification of mineral admixtures (Mehta 1983)

Classification	Chemical and mineral composition	Particle characteristics
II. Highly active pozzolans		
a. Condensed silica fume	Consists essentially of pure silica in noncrystalline form.	Extremely fine powder consisting of solid spheres of 0.1 μm average diameter (about 20 m^2/g surface area by nitrogen adsorption).
b. Rice husk ash (Mehta–Pitt process)	Consists essentially of pure silica in noncrystalline form.	Particles are generally less than 45 μm but they are highly cellular (about 60 m^2/g surface area by nitrogen adsorption.
III. Normal pozzolans		
a. Low-calcium fly ash	Mostly silica glass containing aluminum, iron and alkalies. The small quantity of crystalline matter present consists generally of quartz, mullite, sillimanite, heatite, magnetite.	Powder corresponding to 15–30% particles larger than 45 μm (usually 200–300 m^2/kg Blaine). Most particles are solid spheres with diamter of 20 μm. Cenoshperes and plerospheres may be present.
b. Natural materials	Besides aluminosilicate glass, natural pozzolans contain quartz, feldspar, mica.	Particles are ground to mostly under 45 μm and have rough texture.
IV. Weak pozzolans		
Slow-cooled blast furnace slag, bottom ash, boiler slag, field-burnt rice husk ash.	Consists essentially of crystalline silicate minerals, and only a small amount of noncrystalline matter.	The materials must be pulverized to very fine particle size in order to develop some pozzolanic activity. Ground particles are rough in texture.

TABLE 2.3   Performance-based classification of industrial by-products as mineral admixtures (Mehta 1989)

Class	Description	Example
I	Cementitious	Granulated blast furnace slag
II	Cementitious: pozzolanic	High-calcium fly ash
III	Highly active pozzolans	Silica fume, rice husk ash (Mehta–Pitt process)
IV	Normal pozzolans	Low-calcium fly ash
V	Weak pozzolans	Slow-cooled blast furnace slag, field burnt rice husk ash

Canada. Likewise some of the bituminous coals can produce ash that is not Class F. It is believed that the calcium in the glass affects the pozzolanic activity of fly ash rather than free lime.

Tables 2.2 and 2.3 present performance based classifications of industrial by-products as mineral admixtures for use in concrete devised by Mehta (1983, 1989).

This classification includes high calcium fly ashes as cementitious and pozzolanic materials. But a review of data on chemical and physical properties of fly ashes from different parts of the world suggests that the classification as suggested in Table 2.2 does not represent the correct picture. There are many ashes which have less than 15% CaO content, as per chemical composition, yet such ashes are not only pozzolanic but self cementitious.

Fly ashes produced by four major power plants in Alberta, a Western province of Canada, are such ashes. They contain less than 15% CaO yet are marginally self cementitious and highly pozzolanic. The authors are of the opinion that the degree of self hardening or self cementitious value seems to be related to the coal source rather than the CaO content of the ash.

TABLE 2.4   Proposed classification of fly ash as pozzolanic admixture in cement

Class	Description	Example
I	Pozzolanic but non-self cementitious	Bituminous ash generally exhibiting 2% or more loss on ignition
II	Pozzolanic and self cementitious	Sub-bituminous and lignite ash containing less than 1% loss on ignition

Extensive literature review and experimental study by the authors suggest that the fly ashes generated from bituminous coals with 2 to 6% loss on ignition are invariably pozzolanic but not self cementitious. On the other hand, the fly ashes produced from sub-bituminous and lignite coals invariably contain less than 0.5% loss on ignition and are self cementitious to some degree besides being pozzolanic (Joshi and Lohtia 1995).

Joshi and Lam (1987) and others (EPRI, 1987) have reported that some fly ashes when mixed with water produce heat of hydration and harden like cement. These ashes are generally produced from sub-bituminous and lignite coals from the eastern flats of the Rocky Mountains. Self cementitious ashes are generally finer than the bituminous ashes which are pozzolanic but non-self cementitious. The main difference between the self cementitious and non-self cementitious ashes appears to be that the former have less than 0.5% loss of ignition as compared to 2–6% for the latter. Thus the authors are of the opinion that the fly ashes could be classified not as Class F or Class C, but as low and high ignition loss fly ashes as indicated in Table 2.4 (Joshi and Lohtia 1995).

2.3. FLY ASH CHARACTERISTICS

The effective utilization or disposal of coal ashes requires adequate knowledge of their physical, chemical and mineralogical properties. During the last two decades or so, various investigators have concluded that the chemical composition of an ash has little, if any, effect on its pozzolanic activity. The mineralogical composition, crystalline and noncrystalline phases, particle morphology as well as size and physical make up largely control the pozzolanic reactivity of a fly ash. Because of the inherent variability of fly ash, it is necessary to study the characteristics of a large number of fly ashes so that a large database is generated. Ambitious test programs incorporating more than a thousand fly ashes are underway for preparing such a database by the Western Fly Ash Research and Development and Data Centre (WFARDDC) at the University of North Dakota and North Dakota State University. A similar research project "classification of fly ash for use in cement and concrete" by Electric Power Research Institute (EPRI, 1987) involved extensive physical and chemical characterization of a wide range of fly ash samples from different regions of the United States.

Mathematical models have been used to relate the performance of concrete to the characteristics of representative fly ashes from different regions. Published data suggest that the principal physical characteristics of fly ash which affect concrete performance are particle size, loss on ignition, moisture content (water requirement) and, indirectly, specific gravity (EPRI 1987). The major chemical characteristics affecting properties of concrete include free lime, silicates, aluminates, iron oxide, and carbon. Mineralogy and microstructure including morphology of fly ash are also known to play important roles in the performance of specific fly ashes in cement and concrete.

Coal particles are burned at high temperatures in the furnace where volatile matter is vaporized and the carbon is burned off. However, the inorganic minerallic matter in coal, present in the form of impurities, is converted into ash. Up to 95% of this mineral matter may be composed of clays, pyrite and calcite. During combustion these mineral particles undergo physical and chemical changes in the presence of excess air at high temperatures. As a result several crystalline and glassy phases are formed. The pyrites change to iron oxide, whereas the clay and mica particles slag and partially vitrify to form small glassy spheres composed of amorphous alumino-silicates. Calcination of calcite gives rise to calcium oxide (CaO), calcium hydroxide {Ca(OH$_2$)}, and carbon dioxide (CO$_2$). Intermixed particles of clay and calcite and gaseous matter produce some calcium silicate (CaSiO$_3$), calcium aluminate (CaAl$_2$O$_4$) and calcium sulphate (CaSO$_4$) (Joshi, 1980). Other carbonates and some chlorites, if present in coal, undergo, volatilization and sulphation to produce sulphates, carbon dioxide (CO$_2$) and hydrochloric acid. Quartz particles are rather unaffected and pass through the flame zone without much change in shape.

The thermally altered mineral matter produced from coal combustion in a furnace is quenched as it leaves the flame zone. The fine particulate matter is separated as fly ash from the flue gases by electrostatic precipitators and fabric filters. Due to quenching, spherical to rounded fly ash particles have glassy exterior surface. Some of the gases evolved during combustion are trapped in the fly ash particles, producing low specific gravity cenospheres which float on water and are, therefore, also called floaters, The glass fraction in fly ashes usually varies between 70 and 89% depending on the type and coal source, degree of coal pulverization, combustion conditions in the furnace, and the rate of cooling of the combustion residue.

Fly ash disposal and utilization methods also alter the ash characteristics and affect the properties of the materials in which ash is used. Leachates from ash deposits, compacted fill and concrete may contain some heavy metals. Thus for utilization or disposal of fly ash, the characterization of fly ash for its physical properties, chemistry, microstructure including morphology, and mineralogy is essential. Properties of commonly used fly ashes can also be obtained from the existing data banks on representative fly ashes.

2.4. PHYSICAL PROPERTIES

In the majority of studies conducted for classification and/or characterizations of fly ashes to predict performance of specific fly ash in cement and concrete, physical properties such as particle shape, size and distribution, fineness, specific gravity, and pozzolanic activity index are considered as the main parameters. Some additional tests for quality control of fly ash to meet the ASTM requirements also deal with the determination of autoclave soundness, drying shrinkage, and reactivity with alkalies. A brief description of physical properties of fly ash is presented in this section.

2.4.1. Particle Morphology

In general fly ash particles consist of clear glass spheres and a spongy aggregate. Morphological studies on particle shape and surface characteristics of various types of fly ashes have been conducted by Mehta (1984 and 1988), Diamond (1985), Joshi (1968), Joshi and Lam (1987), and many others, using scanning electron microscope and energy dispersive x-ray analysis (EDXA). Figure 2.1 shows the scanning electron micrographs of different types of fly ash particles. The pictures show the typically spherical shape of the fly ash particles, some of which are hollow. The hollow spherical particles are called cenospheres or floaters as they are very light and tend to float on water surface. Sometimes fly ashes may also contain many small spherical particles within a large glassy sphere, called pherospheres. The exterior surfaces of the solid and hollow spherical particles of low calcium oxide fly ashes are generally smooth and better defined than those of high calcium oxide fly ashes which may have surface coatings of material rich in calcium. In a number of fly ashes from different sources, the

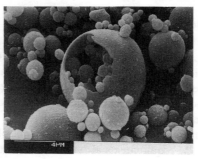

Cenosphere 5000×

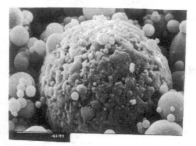

Calcium-rich particle 5000×

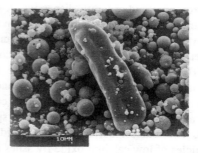

Irregular shaped particle 2000×

FIGURE 2.1   Scanning electron micrographs of different types of fly ash particles.

presence of a significant amount of non-spherical or angular particles has also been detected by several investigators.

2.4.2. Fineness

Fineness is one of the primary physical characteristics of fly ash that relates to its pozzolanic activity (Joshi 1970). Currently ASTM C618 sets the limit for a maximum amount of fly ash retained on the 45 μm (#325) mesh sieve on wet sieving at 34% as quality control measure. A large fraction of ash particle is smaller than 3 μm in size. In bituminous ashes, the particle sizes range from less than 1 to over 100 μm. The average size lies in the range of 7 to 12 μm (Joshi 1968).

Fineness of fly ash is also defined by specific surface area determined by the Blaine air permeability method or by nitrogen adsorption method. Joshi and Marsh (1987) conducted tests for physical properties of 14 Canadian fly ashes. The test results of their studies are presented in Table 2.5. For most Canadian fly ashes, surface area ranges from 1700 to 5900 cm^2/g. Table 2.6 presents a summary of test results for wide range of American fly ashes. The Blaine fineness of American fly ashes varied from 1579 to 5550 cm^2/g. A higher specific surface area can be due to either large amounts of spongy minerallic particles in coal ash or a significant amount of porous carbon particles (Joshi 1968). Several investigators have reported that no consistent relationship between surface area by the Blaine air permeability method and the performance characteristics of fly ash as pozzolans in mortar or concrete exists. As such in the present ASTM specification, no limits are specified on surface area for fineness of fly ash determined by permeability test methods.

Several studies (Mehta 1984, Diamond 1985) suggest that in addition to particle size, gradation is an important performance parameter which can be related to the reactivity of different fly ashes. A large percentage of fly ash particles larger than the 45

TABLE 2.5 Physical properties of 14 Canadian fly ashes (Joshi and Marsh 1987)

Fly ash no.	Coal type	Source	Overall apparent specific gravity	% retained on 45 μm sieve	Specific surface area (m²/g) Air permeability	Specific surface area (m²/g) B.E.T. method
1	Sub-bituminous	Alberta	2.19	32.0	0.42	1.52
2	Sub-bituminous	Alberta	1.92	26.0	0.46	1.61
3	Sub-bituminous	Alberta	1.91	22.0	0.43	1.61
4	Sub-bituminous	Alberta	2.03	9.8	0.59	1.64
5	Lignite	Saskatchewan	2.54	2.8	0.50	1.14
6	Sub-bituminous	Saskatchewan	2.15	20.4	0.22	6.70
7	Lignite	Saskatchewan	2.37	44.8	0.17	1.01
8	Lignite	Saskatchewan	2.39	26.6	0.22	3.47
9	Bituminous	Ontario	2.46	24.0	0.28	0.55
10	Bituminous	Ontario	2.31	27.0	0.25	3.28
11	Bituminous	New Brunswick	2.94	21.4	0.31	2.68
12	Bituminous	New Brunswick	2.87	26.4	0.18	0.43
13	Bituminous	Nova Scotia	2.53	28.2	0.36	0.85
14	Sub-bituminous	Nova Scotia	2.44	34.4	0.38	0.67

Average difference = 1.10 m²/g; standard deviation of mean of 14 readings = 1.44 m²/g.

TABLE 2.6   Analysis of American fly ashes, parameter ranges[a] (EPRI 1987)

Parameter	Range
Silicon dioxide, %	30.92–62.76
Aluminum oxide, %	12.30–26.95
Iron oxide, %	2.84–24.43
Calcium oxide, %	1.10–30.53
Magnesium oxide, %	0.69–6.69
Sulphur trioxide	0.31–3.85
Sodium oxide, %	0.20–2.04
Potassium oxide, %	0.22–3.03
Available alkalies (as equivalent Na_2O), %	0.23–1.54
Loss on ignition, %	<0.01–16.60
Carbon, %	0.02–15.34
Specific gravity	2.14–2.69
Moisture content, %	0.0–0.38
% retained on #325 sieve	3.55–36.90
Blaine fineness (cm^2/g)	
As received	1579–5550
Passed #325 sieve	1804–5350
Pozzolanic activity index, %	
Control #1	86–172
Control #2	136–239

[a]Results of the sampling effort, which included base load and upset condition fly ash samples from 16 pulverized coal-fired power plants across the United States employing electrostatic precipitators, for ash collection.

μm (#325) has been reported to have a negative effect on the 28 day and 90 day strengths of normally cured Portland cement–fly ash mortars. At the same time a large percentage of particles less than 10 μm had a positive influence on mortar strengths (Frigione et al. 1993a). Johsi (1982) also conducted studies on the effect of coarse fraction larger than the 45 μm (#325) on the setting time and strength development of concrete. The data suggested an increase in setting time and only a slight decrease in strength.

The ACI Summary Report 226 (1987) states that the air permeability test procedure provides a quick test for detecting variations in fly ash. Increased surface area, as determined by air permeability tests, in many cases correlates well with quality of a fly ash from a single source.

2.4.3. Specific Gravity

The specific gravity of fly ash is reported to be related to shape, colour as well as chemical composition of fly ash particle. It is adopted as an indirect performance parameter for determining the performance of fly ash in concrete. In ASTM C618, for quality control of fly ash, the uniformity of the fly ash is monitored by limiting the variability of the specific gravity and fineness as measured by the amount retained on 45 μm (#325) mesh sieve. The requirement is that any sample tested shall not deviate from the average of 10 previous tests, or the total of all tests if the number is less than 10, by more than 5%.

In general specific gravity of fly ash may very from 1.3 to 4.8 (Joshi 1968). However, the Canadian fly ashes have specific gravity ranging from 1.91 to 2.94, as can be seen from Table 2.5, whereas from Table 2.6 it is noticed that the American ashes have specific gravity between 2.14 and 2.69. Coal particles with some minerallic impurities have specific gravity between 1.3 and 1.6. Opaque spherical magnetite (ferrite spinel) and hematite particles, light brown to black in colour, when present in sufficient quantity in fly ash increase the specific gravity to about 3.6 to 4.8. As the amount of quartz and mullite increases, the specific gravity decreases. Fly ash pulverization releases some of the gases trapped, during quenching inside the large hollow spherical particles, and increases the bulk specific gravity of the fly ash (Joshi 1968, 1979).

2.4.4. Pozzolanic Activity

Fly ash particles are produced almost in the same way as the ash is produced from a volcano. Both the ashes fall under the group of finely divided admixtures as per ACI committee 212 and exhibit pozzolanic activity. Like volcanic ashes, all commercial fly ashes, even those which possess little or no cementing value, react with calcium oxide in the presence of water and produce highly cementitious water insoluble products. This property of the ashes is known as pozzolanic reactivity. The meta-stable silicates present in self cementitious fly ash react with calcium ions in the presence of moisture to form water insoluble calcium–alumino-silicate hydrates.

The pozzolanic activity of a fly ash depends upon many parameters, most important of which are fineness, amorphous matter, chemical and mineralogical composition and the unburned carbon content or loss on ignition of the fly ash (Joshi 1979). Several investigators have reported that when fly ash is pulverized to increase fineness, its pozzolanic activity increases significantly. However, the effect of increase in specific surface area beyond 6000 cm^2/g is reported to be insignificant (Joshi and Marsh 1987).

Chemical composition of fly ash does not reflect the form in which various compounds are present. Yet the two governing parameters indicative of the reactivity are its calcium oxide and carbon content. Sub-bituminous and lignite Class C fly ashes with high content of CaO exhibit self hardening property besides pozzolanic activity. Some of the Class C fly ashes generate heat of hydration immediately on adding water.

Carbon content or unburned carbon in fly ash is generally determined by the standard loss on ignition test. Carbon also acts as diluent of the active pozzolanic matter in the fly ash and contributes to size fractions larger than 45 μm. Because of the undesirable effects of carbon on the pozzolanic activity, different organizations specifying the use of fly ash as pozzolan limit the loss on ignition in fly ash from 3 to 6%. In Class C fly ashes, loss on ignition even in excess of 0.5% seems to affect the pozzolanic activity as well as some other performance characteristics significantly (Joshi and Lohtia 1995). The effect of iron oxide on the pozzolanic activity of fly ash is not that significant when high amount of silica is present.

Pozzolanic activity of fly ash is determined in terms of strength activity index. The current ASTM requirement for pozzolanic activity with cement is that the strength developed by the specimens of the test mixture, in which 35% of the cement by weight is replaced by the fly ash being tested, shall be a minimum of 75% of the strength of the control specimens after storage at $38 \pm 1.7°C$ ($100 \pm 3°F$) for one day and then at $55 \pm 1.7°C$ ($131 \pm 3°F$) for six days. This is similar to the pozzolanic activity requirement of the American Association of State Highway and Transportation Official (AASHTO) specification M295. Under the proposed changes to ASTM C618, the fly ash would be acceptable on the basis of the 7 day test, but would be rejected only if it failed the 28 day test as described in ASTM C311. Most fly ashes meet the 7 day limits. As a result, the time for evaluating their pozzolanic activity or suitability as a pozzolan as per new test procedure would be considerably reduced.

All the tests for pozzolanic activity as described in various specifications require additional tests to ensure durability criteria. Although initially intended to be a method of the ability of the fly ash to develop strength by pozzolanic reactions, it has been reported that the quality of lime or cement used in the test significantly affects results. The fly ashes may pass the tests with one cement or lime and fail with others. In addition, there appears to be no direct and consistent relationship of the results of the tests to the long term performance of specific fly ashes in cement concrete.

A simple test for evaluating pozzolanic activity of fly ashes and other pozzolans has so far evaded the efforts of the scientific community. Potential users of fly ash do not seem to have full confidence in the current ASTM classification guideline test, because of its limited predictive capabilities of the performance of the fly ash in concrete. For increased use of the fly ash in concrete, alternatives to the present pozzolanic activity test have been proposed by several researchers (Mehta 1984, Diamond 1985, Dodson 1985). A measure of activity is being sought by the Electric Power Research Institute (EPRI). Research is currently being conducted, but no general agreement for a better or quick test for evaluation of pozzolanic activity has yet been reached.

Detailed investigations by Joshi and Rosauer (1973a,b) on synthetic ashes, produced from minerals associated with coals in laboratory scale furnace, suggest that strain in glass is the main source for pozzolanic activity. The fineness and chemical composition of course dictate the degree of pozzolanicity in a fly ash containing strained glass. But measurement of strain in glass particles in fly ash has proven to be an uphill task. The importance of strain in glass has become more apparent since the introduction of low NO_x burners in power plants using suspension-fired furnaces. The flame temperature in such furnaces is much lower than the normal suspension-fired furnaces. Also the ash particle residence time in the low NO_x furnace is much longer. Therefore, the thermal conditions are quite different. The ash produced in such power plants is sometimes not sufficiently pozzolanic and therefore is considered unsuitable for use as a pozzolan in cement and concrete.

TABLE 2.7 Average chemical analyses of fly ashes from various countries

Country		Number comp. sample	Chemical Composition (weight per cent)									
			Loss on ignition	SiO$_2$	Al$_2$O$_3$	Fe$_2$O$_3$	CaO	MgO	SO$_3$	Na$_2$O	K$_2$O	Total
Japan	Av	12	0.73	57.96	25.86	4.31	3.98	1.58	0.34	1.49	2.15	8.40
	Max		1.23	63.27	28.35	5.90	6.74	2.09	0.81	2.36	3.15	99.27
	Min		0.06	53.41	22.88	2.82	1.04	1.00	0.02	0.88	1.73	97.48
	R		1.07	9.86	5.47	3.08	5.07	1.09	0.79	1.48	1.42	1.79
	s		0.36	2.94	1.32	0.81	1.64	0.44	0.26	0.44	1.38	0.57
USA	Av	34	7.83	44.11	20.81	17.49	4.75	1.12	1.19	0.73	1.97	99.73
	Max		18.00	51.09	28.03	31.30	12.00	1.40	2.80	2.10	2.98	100.55
	Min		1.00	32.70	14.60	8.50	11.10	0.70	0.30	0.22	1.28	97.94
	R		17.00	19.20	13.70	22.80	0.90	0.70	2.50	1.88	1.70	2.61
	s		4.75	4.52	3.67	6.07	2.91	0.33	0.79	0.51	0.46	0.53
Great Britain	Av	14	3.86	46.16	26.99	10.44	3.06	1.99	1.59	0.90	3.26	98.22
	Max		11.70	50.70	34.10	13.50	7.70	2.90	6.80	1.90	4.20	102.90
	Min		0.60	41.40	23.90	6.40	1.70	1.40	0.60	0.20	1.80	96.10
	R		11.10	9.30	10.20	7.10	6.00	1.50	6.20	1.70	2.40	6.80
	s		3.62	2.53	2.50	2.11	1.49	0.41	1.58	0.51	0.72	1.60

TABLE 2.7 continued

Country		n								n=15	n=15	100.1 / 3
France	Av	17	3.72	48.45	25.89	8.70	5.95	2.36	1.01	0.64	3.94	100.1
	Max		15.15	54.05	33.40	15.30	38.75	4.45	7.00	0.85	6.00	—
	Min		0.30	29.90	10.80	5.80	1.48	1.10	0.10	0.15	0.70	—
	R		14.85	24.15	22.60	9.50	37.27	3.35	6.90	0.70	5.30	—
	s		4.01	5.81	6.03	2.20	9.31	0.91	1.73	0.19	1.23	—
Germany	Av	9	9.65	41.13	24.39	13.93	5.06	1.83	0.77	—	—	96.78
	Max		20.10	49.54	29.35	20.88	11.81	4.26	2.10	—	—	98.35
	Min		1.48	34.10	21.06	8.37	2.18	0.75	0.12	—	—	94.33
	R		18.62	15.44	8.29	12.51	9.63	3.51	1.98	—	—	4.02
	s		7.14	5.38	3.43	4.64	3.37	1.15	0.62	—	—	1.25
USSR	Av	15	—	55.08	25.97	7.83	5.08	1.81	1.63 (n=11)	—	—	97.40
	Max		—	62.08	37.15	12.01	10.62	2.90	3.78	—	—	—
	Min		—	47.90	20.71	3.08	1.10	0.28	0.20	—	—	—
	R		—	14.18	16.44	8.93	9.52	2.62	3.58	—	—	—
	s		—	5.20	4.57	2.49	2.77	0.79	1.39	—	—	—

R = range, s = sample standard deviation.

2.5. CHEMICAL PROPERTIES

Chemical constituents in fly ash reported in terms of oxides include silica (SiO_2), alumina (Al_2O_3), and oxides of calcium (CaO), iron (Fe_2O_3), magnesium (MgO), titanium or vertiel (TiO_2), sulphur (SO_3), sodium (Na_2O), and potassium (K_2O). Unburned carbon is another major constituent in all the ashes. Amongst these SiO_2 and Al_2O_3 together make up about 40 to 80% of the total ash. The sub-bituminous and lignite coal ashes have a relatively higher proportion of CaO and MgO and lesser proportions of SiO_2, Al_2O_3 and Fe_2O_3 as compared to the bituminous coal ashes.

Range of various constituents in the fly ashes on a worldwide basis is presented in Table 2.7. The test results of chemical properties of 14 Canadian fly ashes are given in Table 2.8. Table 2.6 includes the range of chemical and physical parameters based upon analyses of a large number of fly ashes collected from thermal power plants all over the United States.

It is observed that fly ashes from sub-bituminous and lignite coals contain relatively large proportion of calcium oxide (CaO) and Magnesium oxide. Such ashes are characterized as Class C ashes as per ASTM C618. On the other hand, the ashes produced from bituminous coal are relatively rich in ferric oxide (Fe_2O_3) and contain less than 5% calcium oxide. Table 2.9 presents ranges of CaO content of North American fly ashes derived from various classes of coal. Canadian fly ashes have CaO ranging from 0.75 to 20.5% compared to 1.1–30% in American fly ashes.

Bituminous coals produce fly ashes with relatively higher amounts of unburned carbon, reported as percent loss on ignition (LOI), than sub-bituminous or lignite coals. The amount of unburned carbon also depends to some extent upon the degree of coal pulverization, rate of combustion, and air/fuel ratio in addition to the type and source of coal. ASTM C618 includes the chemical requirements related to the suitability for use of

TABLE 2.8 Summary of test data on chemical composition of 14 Canadian fly ashes (Joshi and Marsh 1987)

Fly ash[a] No.	Weight percent ± 1σ											
	Al as Al_2O_3	Fe as Fe_2O_3	Ti as TiO_2	Ca as CaO	Mg as MgO	S as SO_3	Na as Na_2O	K as K_2O	Si as SiO_2	LOI %	Magnetic portion[b]	Water soluble fraction wt%
1	22.3±0.5	4.1±0.1	1.01±0.01	13.6±0.4	1.58±0.04	0.2±0.1	0.27±0.03	1.02±0.06	54.8±0.4	0.26	≤1	4±0.5
2	23.1±0.5	3.7±0.1	0.78±0.01	13.0±0.4	1.17±0.04	0.2±0.1	2.19±0.03	1.26±0.06	57.6±0.4	0.15	≤1	2±0.5
3	21.0±0.5	3.5±0.1	0.50±0.01	11.6±0.4	0.71±0.04	≤0.1	1.68±0.03	1.51±0.06	56.4±0.4	0.61	≤1	2±0.5
4	22.7±0.5	3.7±0.1	0.60±0.01	8.4±0.4	0.75±0.04	≤0.1	2.36±0.03	1.02±0.06	57.4±0.4	0.10	≤1	4.5±0.5
5	23.7±0.5	4.4±0.1	0.94±0.01	19.2±0.4	3.96±0.04	0.6±0.1	0.54±0.03	2.41±0.06	57.9±0.4	0.13	≤1	7±0.5
6	20.4±0.5	5.4±0.1	0.60±0.01	12.2±0.4	1.38±0.04	0.1±0.1	0.84±0.3	1.93±0.06	52.6±0.4	0.12	≤1	1±0.5
7	20.8±0.5	4.1±0.1	0.97±0.01	20.0±0.4	3.13±0.04	0.2±0.1	5.66±0.07	1.14±0.06	40.7±0.4	0.83	≤1	2.5±0.5
8	20.8±0.5	3.8±0.1	0.87±0.01	18.6±0.4	2.42±0.04	≤0.1	5.80±0.07	1.45±0.06	38.1±0.4	0.34	≤1	3.5±0.5
9	22.7±0.5	16.5±0.8	1.01±0.01	4.6±0.4	0.64±0.04	0.5±0.1	0.54±0.03	2.0±0.06	44.6±0.5	5.9	85.±1.	3.5±0.5
10	21.2±0.5	7.1±0.8	0.81±0.01	6.6±0.4	0.88±0.04	0.1±0.1	0.4±0.03	1.14±0.06	51.9±0.5	5.3	69.±1.	2.5±0.5
11	13.6±0.5	38.5±0.8	0.51±0.01	1.81±0.04	0.14±0.01	0.6±0.1	0.11±0.01	1.93±0.06	34.9±0.4	0.11	96.±1.	1.5±0.5
12	13.6±0.5	42.2±0.8	0.56±0.01	1.22±0.04	0.06±0.01	0.4±0.1	0.09±0.01	1.26±0.06	31.7±0.4	2.0	87.±1.	1±0.5
13	20.8±0.5	24.4±0.8	0.76±0.01	0.94±0.04	0.12±0.01	0.6±0.1	0.34±0.03	1.63±0.06	46.7±0.4	0.53	94.±1.	1.5±0.5
14	18.0±0.5	19.2±0.2	0.74±0.01	0.76±0.04	0.75±0.04	0.99±0.05	0.39±0.03	2.93±0.06	44.7±0.4	6.25		1±0.5

[a]See Table 2.5 for fly ash source. [b]This is the amount in the sample attracted to a magnet. It does not indicate the amount of magnetite fraction in ash.

TABLE 2.9   Range of CaO content in North
American fly ashes

Type of coal	CaO content (%) in fly ash
Eastern bituminous	1–6
Colorado bituminous	4–8
Utah and Alberta sub-bituminous	6–12
Texas lignite	7–12
Saskatchewan lignite	10–15
North Dakota lignite	18–25
Montana and Wyoming sub-bituminous	22–32

fly ash as mineral admixture in cement concrete as shown in Table 2.1.

2.5.1. Total Oxides

For Class F fly ashes, the sum of silica, alumina and iron oxide ($SiO_2 + Al_2O_3 + Fe_2O_3$), must be at least 70.0%, whereas for Class C the required minimum is 50.0%. For fly ash to act as a pozzolan, it is necessary to have chemical constituents capable of reacting with lime in the presence of water. Thus a minimum proportion of total SiO_2 plus Al_2O_3 plus Fe_2O_3 is specified to ensure that sufficient reactive constituents are present. The lower requirement for Class C fly ashes recognizes that considerable amount of CaO will be present in self hardening cementitious materials and thus the percentage of the pozzolanic components may therefore be lower. As discussed later, only the glassy or amorphous phases of oxides take part in pozzolanic reaction. Studies by Mehta (1984), Joshi (1970) and many others have shown no definitive relationships between pozzo-lanic activity and proportion of individual oxides as such.

2.5.2. Sulphur Trioxide (SO₃)

The maximum SO_3 content allowed in the fly ash by ASTM C618 is 5.0%. The SO_3 content has been reported to affect to some degree the early age compressive strength of mortar and concrete specimens. The higher the SO_3 content, the higher is the resultant strengths. Different cements require different amounts of SO_3 for the development of maximum strength. Generally the SO_3 limits are set below the optimum limit and thus the added SO_3 from fly ash may be an advantage in some cases. However, a maximum limit on SO_3 is considered necessary to avoid an excess sulphate in hardened concrete which may contribute to disruptive sulphate attack.

2.5.3. Moisture

A 3.0% limit on moisture content is specified in ASTM C618 to prevent caking and packing of the fly ash during shipping and storage and to control uniformity of fly ash shipments. This specification will also avoid sale and handling of significant amounts of water as a part of the admixture. Any amount of moisture in Class C fly ash will cause hardening from hydration of its cementitious compounds. Even surface spraying may cause caking. Therefore such ashes have to be kept dry before being mixed with cement. As shown in Table 2.6, American fly ashes have moisture content ranging from 0.0 to 0.38%, much below the specified limits.

2.5.4. Carbon Content (Loss on Ignition)

The need for a low carbon content in fly ash is related to requirements for proper air entrainment. More air entraining agent is required to entrain a specified amount of air in fly ash concrete than is required for a similar concrete without fly ash.

The maximum permissible loss on ignition is related to the amount of carbon or unburned coal in the fly ash. For some of the fly ashes there is a significant difference in carbon content and loss on ignition (LOI). Yet the specification restricts the LOI and not the unburned carbon. The present ASTM C618 sets the limit on loss on ignition at 6.0% for Class F and Class C fly ashes. For the use of fly ash concrete in highway construction in the United States, The AASHTO specification M295 places the LOI limit at 5.0%. Table 2.8 shows that Canadian fly ashes have LOI form less than 1.0% to greater than 6%. The CaO content of Canadian fly ashes varies from 0.76 to 20.0%.

Erratic entrainment of air has been encountered with many fly ashes containing carbon contents even less than 1.0%. This has hampered the increased use of fly ash in air entrained concrete because of quality assurance of fly ash concrete as well as cost considerations.

Fly ash is normally finer than cement and also has specific gravity less than cement. As a result the volume of fly ash added is usually greater than the volume of cement replaced. Because of this, surface area of the binder within the concrete mix is increased. Thus greater volume of air entraining agent is needed to provide the same concentrations of the air voids in the mortar or concrete containing fly ash.

The second probable reason for increased amount of air entraining agent is related to the type and amount of carbon in the fly ash. The carbon absorbs a portion of the air entraining agent, which limits its availability for producing the needed stable air bubbles. The amount of adsorption varies with the amount of carbon present and possibly, with the form of such carbon. Thus variations in the carbon content result in a variation in air entraining agent demand. The presence of organic constituents other than carbon may interact with the air entraining agent to reduce its effectiveness. The presence of activated carbon may also affect the effectiveness of other admixtures used in concrete.

Although there appears to be no evidence that high carbon content in fly ash is detrimental to the strength development and durability of fly ash concrete when appropriate air entrainment is attained, the problems associated with obtaining the proper air entrainment may over ride all potential advantages of using fly ash in concrete. Thus the variable and erratic behaviour with some combinations of ingredients will permit inadequate or excess air in some portions of a structure to go undetected. Such variations make the use of fly ash concrete questionable in the absence if any other advantages derived from the properties of fly ash concrete.

Under the present conditions, many power companies using pulverized coal as fuel are concerned with achieving maximum efficiency for burning as well as producing high quality fly ash with less carbon content. Research studies to investigate the phenomena that control the amount of air entraining agent for a given amount of air in the fly ash concrete are underway to permit unhindered and increased usage of fly ash in concrete.

Of course, loss on ignition test as per ASTM C311 is still the test most often used to determine the carbon content in fly ash. However, equipment for a rapid direct determination of carbon is now employed by a number of power utility companies. An example of such equipment is the LECO combustion furnace that automatically indicates the carbon content in ash.

2.5.5. Magnesium Oxide

Earlier versions of ASTM C618 contained an optional require-ment permitting the maximum limit on magnesium oxide (MgO) of 5.0% in fly ash. This has been now dropped. The purpose of initially including this requirement was to control expansion of the fly ash concrete. The presence of MgO in a form capable of hydrating to form magnesium hydroxide (brucite) in the hardened concrete is suspected to cause disruptive expansion. In the current ASTM standard, autoclave expansion test is employed to ensure proper soundness and to avoid disruptive expansion of concrete containing fly ash. The optional requirement is, however, still retained in AASHTO M295 and several other standards.

2.5.6. Available Alkalies

Available Alkalies in a fly ash are determined after an intimate mixture of lime, fly ash and water has been stored for 28 days at 37.8°C. The available alkalies refer to those soluble in hot water after the period of storage. Both ASTM C618 and AASHTO M295 permit a maximum of 1.5% of available alkalies. This requirement is needed only where the aggregates susceptible to alkali aggregate reaction are encountered.

The proportion of total alkalies that become water soluble when the fly ash is mixed with lime and water is dependent on the temperature and duration of storage. Thus available alkalies determined by the test may not represent the field conditions. However, the maximum limit offers protection against undesirable amounts of sodium and potassium ions in the fly ash concrete. The test is under revision to reflect the field conditions of curing and placement of concrete.

2.6. MINERALOGICAL CHARACTERISTICS

An x-ray diffraction study of the crystalline and glassy phases of a fly ash is commonly called mineralogical analysis of fly ash. Fly ashes, generally, have 15 to 45% crystalline matter. The high calcium ashes, derived from low-rank sub-bituminous and lignite coals, contain larger amounts of crystalline matter ranging between 25 and 45%. Although high calcium Class C ashes may have somewhat less glassy or amorphous material, they also have certain crystalline phases such as anhydride ($CaSO_4$), tricalcium aluminate ($3CaOAl_2O_3$), calcium sulpho-aluminate ($CaSAl_2O_3$) and very small amount of free lime (CaO) that participate in producing cementitious compounds. Furthermore glassy phase in Class C ashes is usually considered more reactive. The glassy particles in Class C fly ashes seem to contain large amount of calcium which possibly also makes the surface of such particles highly strained. It is possibly for this reason that the Class C fly ashes are highly reactive.

Table 2.10 presents crystalline phases in North American fly ashes identified by XRD analysis. The crystalline phases can be identified using x-ray diffraction (XRD), energy dispersive x-ray analysis (EDXA), and scanning electron microscopic studies. The fly ash mineralogy presented in Table 2.10 was determined by a semi-quantitative x-ray powder diffraction analysis.

It is reported that most of the reactive portion of fly ash is on the surface of the particles where the influence of quenching is maximum. Also, as the fly ash particle size increases, the amount of crystalline silica (SiO_2) in the ash increases and the proportion of calcium containing compounds decreases. Therefore, large size particles are less reactive.

Qualitative XRD data presented in Table 2.10 indicate that low calcium Class F fly ash consists typically of the crystalline phases of quartz, mullite, hematite and magnetite in a matrix of alumino-silicate glass. High calcium Class C fly ashes on the other hand, have a much more complex assemblage of crystalline phases that typically contain the four phases present in Class F ashes plus several other phases as given in Table 2.10. For Class C ashes, glass composition among the particles is more heterogeneous and range from calcium aluminate to sodium calcium alumino-silicate.

In order to obtain generic information relevant to the utilization of fly ash in concrete, it is necessary to have thorough understanding of the chemical, physical, and mineralogical characteristics of the available and likely to be produced fly

TABLE 2.10   Crystalline phases in North American fly ash
identified by XRD analysis (McCarthy et al. 1988)

Class of fly ash and code	Name	Nominal composition
Low-calcium/class F		
Hm	Hematite	Fe_2O_3
Mu	Mullite	$Al_6Si_2O_{13}$
Qz	Quartz	SiO_2
Sp	Ferrite spinel	$(Mg,Fe)(Fe,Al)_2O_4$
High-calcium/class C		
Ah	Anhydrite	$CaSO_4$
AS	Alkali sulphate	$(Na,K)_2SO_4$
C_2S	Dicalcium silicate	Ca_2SiO_4
C_3A	Tricalcium aluminate	$Ca_3Al_2O_6$
Hm	Hematite	Fe_2O_3
Lm	Lime	CaO
Ml	Melilite	$Ca_2(Mg,Al)(Al,Si)_2O_7$
Mu	Mullite	$Al_6Si_2O_{13}$
Mw	Merwinite	$Ca_3Mg(SiO_4)_2$
Pc	Periclase	MgO
Qz	Quartz	SiO_2
So	Sodalite structure	$Ca_2(Ca,Na)_6(Al,Si)_{12}O_{24}(SO_4)_{1-2}$
Sp	Ferrite spinel	$(Mg,Fe)(Fe,Al)_2O_4$

ashes in thermal power plants located in a particular region.
Extensive studies in this regard are continually underway for
North American fly ashes with the distinct goal of generating a
large data base.

Mineralogical characterization is valuable in determining the
crystalline phases that contain the major constituents of fly ash,
"the element speciation." Fly ash mineralogy is mostly depen-
dant on the type and composition of the source coal. Calcium

oxide (CaO) content which has been recognized to govern the performance characteristics of fly ash in concrete is dependent on coal mineralogy, and on reactions during combustion. It is generally believed that the pozzolanic activity of the fly ash is influenced by calcium in the glass, and not by the free calcium oxide or crystalline CaO (Joshi 1970). The most variable phases identified in fly ash are anhydrite ($CaSO_4$), periclase (MgO), ferrite spinel (magnetite) and hematite, melinite and lime (CaO). Anhydrite and the nonoxides are largely controlled by the hydrite content of the coal. It is reported that melinite forms from crystallization or devitrification, which is dependent on ash cooling rate in the furnace and flue, and thus could reflect variable furnace operating conditions. Tricalcium aluminate ($3CaOAl_2O_3$) has been observed in most of Class C and some intermediate calcium ashes as well. The role of some of the important phases observed on the properties of fly ash is described below.

2.6.1. Anhydrite ($CaSO_4$)

It forms from reaction of CaO, SO_2, and O_2 in the furnace or flue. The amount of anhydrite increases with increasing SO_3 and CaO contents in the ash. Anhydrite is a characteristic phase in high calcium Class C fly ashes. It has also been detected in some low calcium Class F fly ashes as well as in most of the intermediate calcium ashes. For most ashes, only about half of the SO_3 is present as anhydrite. In case of high SO_3 ashes, alkali, sulphates and calcium sulpho-aluminate are also identified as crystalline phases.

Anhydrite plays a significant role in fly ash hydration behaviour because it participates along with tricalcium aluminate and other soluble aluminates to produce ettringite, calcium sulpho-aluminate hydrate.

The formation of ettringite immediately on adding water to fly ash contributes significantly to the self hardening characteristics of fly ash. Ettringite may also precipitate and control the solubility/leachability of potentially hazardous trace elements from the fly ash thus affecting the geochemical behaviour of the disposed ash in landfills or disposal ponds.

2.6.2. Periclase (MgO)

Periclase refers to the crystalline form of magnesium oxide (MgO). It is always present in high calcium ash and commonly found in intermediate calcium ash. This form of MgO present in fly ash affects the soundness of the resulting concrete through its expansive hydration to brucite, $Mg(OH_2)$. However, the studies have shown that for the ashes where periclase was detected, only half of the MgO was periclase and as such no adverse effects of higher MgO content in fly ash on the soundness of the resulting concrete was observed. Therefore, the requirement of MgO content not exceeding 5.0% which existed until recently in ASTM C618 has been currently dropped.

2.6.3. Ferrite Spinal (Magnetite) and Hematite

Crystalline iron oxide, ferrite spinel and/or hematite has been observed in all the fly ashes. For most ashes only one third to one half of the iron is present as crystalline oxide. The reactivity of a fly ash is however dependent on the glassy phases of Fe_2O_3. There is at least a small amount, from 0.1 to 1%, of iron present as hematite in almost all the fly ashes. High calcium Class C fly ashes have however less amount of hematite as well as total Fe_2O_3. Ferrite spinel refers to the ferrimagnetic spinel structure phase in fly ash which manifests the common tendency for solid substitution of Al, Mg and Ti for Fe.

In the ASTM C618, the pozzolanic activity of a fly ash is assumed to be related to the chemical composition as is evident from the limitations on the sum of SiO_2 and Al_2O_3 present as non-reactive quartz, mullite and other silicates and alumino-silicates. Only the glassy phases of SiO_2 and Al_2O_3 are reported to be pozzolanic as far as fly ash reactivity in concrete is concerned. The more glassy oxides an ash contains, the greater is its potential for pozzolanic activity in cement and concrete, although the composition and structure of the glassy phases are also important.

2.6.4. Tricalcium Aluminate ($3CaOAl_2O_3$)

High calcium Class C fly ashes invariably contain tricalcium aluminate with its relative content increasing with an increase of CaO content of the ash. Sometimes intermediate calcium ashes, with CaO content of 8 to 15%, have also been found to contain this compound. Tricalcium aluminate is one of the most important crystalline phases to identify and quantify in fly ash because it contributes to ettringite formation, and also in self hardening reactions as well as disruptive sulphate reactions in hardened concrete.

Different x-ray diffraction studies indicate that fly ash mineralogy varies significantly, particularly in high calcium Class C fly ashes. Each fly ash is unique and may itself vary with time due to the differences in the chemical composition and mineralogy of the source coal as well as due to the combustion conditions in the furnace. For statistical analysis and to develop predictive models for performance of a fly ash in cement and concrete, a broad data base of the fly ash mineralogy for American fly ashes based on XRD studies and also of their physical and chemical characteristics is under preparation by the Western Fly Ash Research and Development and Data Centre at the University of North Dakota and North Dakota State University (McCarthy and Solem 1991).

2.7. STANDARDS AND SPECIFICATIONS

The world over, fly ash is currently the most abundant and commonly used pozzolan in cement mortar/concrete. Because of the reported difficulty in evaluating degree of pozzolanic activity, most organizations do not even identify fly ash as pozzolanic admixture but as mineral admixture in cement concrete. Therefore, various organizations around the world and even different organizations in the United States have their own standards, codes of practice, guidelines, and requirements for the use of mineral admixtures in cement and concrete. The main object of the standards is to achieve specified performance and provide tests for the characterization and quality control/assurance of the fly ash to alleviate problems associated with the intrinsic variability of natural materials as well as industrial by-products for use as mineral admixtures in cement concrete.

In North America through the co-operative efforts of users, producers and general interest participants sponsored by ASTM Committee C-9 on concrete, ASTM C350 describing specifications for fly ash was first adopted in 1954. In 1968, this was combined with ASTM C402 to form the present ASTM C618 which has been constantly reviewed and modified to keep it current with continual developments and improved test methods. A summary of ASTM C618-93 is presented in Table 2.1. Table 2.11 gives the relevant standards of some countries worldwide. The AASHTO sub committee on materials has set up its specification M295, which is very similar but not identical to ASTM C618. Each of these specifications is for pozzolans rather than for fly ash exclusively and each defines three classes of pozzolans: Class N for natural pozzolans, Class F for low calcium fly ashes, and Class C for high calcium fly ashes. ASTM provides opportunity for input by all groups in developing and modifying the standards for use of mineral admixtures in cement concrete. These include fly ash producers/marketers, concrete producers, concrete users such as transportation

TABLE 2.11   Various standard specifications for
fly ash use as pozzolan

Country	Designation of standard
Australia	AS 1129-1971[a]
Canada	CAN/CSA-A23.5-M86
India	IS 1344-1968[a]
Japan	JIS A6201-1977[a]
Germany	DIN 1045/lfBT[a]
Great Britain	BS 3892 Part 1 and Part 2 (revisions 1982 and 1991)
United States	ASTM C618-1995
USSR	GOST 6269-1963[a]

[a]Most recent revisions not known.

agencies, and academic and other researchers in various government laboratories.

The first British Standard BS 3892 governing the use of fly ash in concrete was introduced in 1965. It was revised in 1982 and the latest version is BS 3892 Part I, 1991 titled "Specifications for pulverized fuel ash for use as a cementitious component for use with Portland cement." BS 6610, 1985 "Specifications for pozzolanic cement with pulverized fuel ash as a pozzolan" refers to the factory produced blended cement using fly ash. Likewise most countries in the world have developed their own standards for test procedures and specification requirements for use of their fly ash as a pozzolan in cement and concrete. Chemical composition, as well as physical and mineralogical characteristics of fly ash are generally the governing factors in one form or another in defining the specification requirements in the majority of the standards. The standards on a global basis are constantly reviewed to accommodate latest developments and test methods.

2.8. QUALITY ASSURANCE PROCEDURES FOR FLY ASH

When fly ash is to be used as a pozzolan in cement concrete, it must act as an active ingredient of concrete. Therefore, considerably more testing and evaluation of fly ash properties are necessary while using the ash in concrete as compared to other uses of the fly ash such as for fills and embankments or as a filler is asphaltic concrete. The inherent variability of fly ash is the major obstacle to continued and increased utilization of fly ash, as well as a problem in modelling the geochemical behaviour of fly ash. This is due to the differences in the inorganic constituents of the source coal, the degree of coal pulverization, combustion conditions in the furnace, and in ash collection and handling methods. Because no two power plants have all of these factors in common, fly ash produced by each plant is unique and may vary with time even in the same plant.

Over the years, the increased use of fly ash and adoption of special equipment and procedures in power plants by power utility companies has made it possible to maintain the ash quality and uniformity as much as practical. The sale of high quality fly ash can be a significant positive economic factor to a power company. In the suspension-fired furnaces the chemical composition of the inorganic fraction of the fly ash is not likely to vary significantly, as long as the same coal is used as fuel in the plant and no start up fuel oil or extraneous matter, such as lime or sodium carbonate, is added. Problems could arise if different coals or different blends of several coals are used. Likewise chemicals are also added to increase collection efficiency of the electrostatic precipitators. Loss on ignition or carbon content and fineness of fly ash are somewhat dependent on the combustion conditions and how well the collectors function. More variability in fly ash is generally caused by additives and furnace conditions than by the inorganic chemical composition of the mineral matter in coal. Most agencies processing fly ash for sale in compliance with ASTM C618 requirements monitor loss on ignition and fineness on a frequent basis, often daily to assure fly ash uniformity and quality.

The standard procedures for sampling and testing fly ash have been developed in many countries for quality control and assurance of this material for use in cement and concrete. In North America, ASTM C311 procedures for "sampling and testing of fly ash or natural pozzolans for use as a mineral admixtures in concrete" recommend that tests for fineness (wet sieving), moisture, specific gravity, loss on ignition, and soundness be conducted for every 400 tonnes of the material. Other tests including chemical analysis, and 28 day cement pozzolanic activity or strength index, are made for each 2,000 tonnes. The sample used for the 2,000 tonne test is made up of a composite of 5 previously tested samples representing 400 tonnes each.

Generally a source of fly ash requires prequalification and approval by the user/construction agencies. Initial approval maybe on the basis of the purchaser's test data or the data provided by the fly ash producer/marketer through an independent approved testing laboratory. The product is then often accepted by certification of compliance with random or periodic check tests described in ASTM C311. Some agencies may require that tests such as fineness and loss on ignition be made on each shipment of fly ash. Random checks on prequalified sources are generally at the discretion of the user or construction agency. How frequently they are made may depend somewhat on the volume of fly ash being used and the history of product from a particular source. All new sources of fly ash are required to be carefully evaluated. Should any change in source of materials be necessary during the course of a construction project, the effects of that change on the characteristics of the resultant fly ash concrete should be evaluated before use of fly ash in concrete.

Fundamental studies involving techniques such as x-ray diffraction (XRD), energy dispersive x-ray analysis (EDXA), scanning election microscopy (SEM), and thermal gravimetric analysis (TGA) as well as differentiated thermal analysis (DTA) have been employed to identify composition and mineralogy of the fly ash and reaction products as well as particle size distribution and morphology. These techniques provide comprehensive evaluation of fly ash characteristics related to their use in cement concrete. In future understanding of the fly ash–cement–water interactions should allow clear identification of the potential for new sources of fly ash and more meaningful pozzolanic activity test than those now included in ASTM C618 and AASHTO M295.

Quality assurance usually includes both the quality control of the fly ash, which is the responsibility of the fly ash producer/marketer, and the acceptance tests and procedures that are the responsibility of the purchasing/user agency. Quality assurance also applies to the use of the fly ash to produce a concrete with specified characteristics both in its fresh and hardened state.

CHAPTER 3

Use of Fly Ash in Cement and Concrete

Fly ash may be used in concrete as a raw material for cement production, as an ingredient in blended cement, and as a partial replacement for cement in concrete. Sometimes fly ash is also used as a partial replacement of fine aggregate as well as in the production of light weight aggregate for concrete. The present state of the art is well established with respect to the use of fly ash as cementitious component or mineral admixture in concrete. Since about 1980, after the oil embargo, and the switch from oil to coal as the major fuel for electric power generation, there has been substantial increase in the amount of fly ash produced in the North American Continent. With the passage of Resource Conservation Recovery Act (RCRA) and introduction of EPA guidelines for the use of fly ash in federal concrete procurement in 1983–84, there has been a surge of research effort with respect to fly ash and its characteristics with considerable emphasis on its use as an ingredient in concrete. To date, most use and, therefore, most research has been for fly ash use in concrete. Coal consumption for power production has increased in the developing countries of Asia in particular. As a result, worldwide there has been annual increase in the production and utilization of fly ash in concrete including first time production and utilization in several countries (Manz 1993).

Generally, fly ash use in concrete has provided cost savings, improved workability and pumpability, better surface finish, lower heat of hydration, improved long term or ultimate strength, reduced permeability, improved sulphate resistance, and reinforcement corrosion prevention. On the debit side, fly ash use in concrete results in delayed strength gain, increased demand of air entraining agent with increasing carbon content in the fly ash, and slightly reduced resistance to scaling due to salts used for deicing on concrete roads.

Fly ash has been used as an admixture or as a primary constituent in both cast-in-place and precast products for more than 50 years. Most use of fly ash in concrete is attributed to its fineness as well as pozzolanic and sometimes self cementitious characteristics.

3.1. BASIC CONCEPT OF FLY ASH AS A POZZOLAN IN CONCRETE

Setting or hardening of ordinary Portland cement concrete occurs due to hydration reaction between water and the cementitious compounds in cement which give rise to several types of hydrates of calcium silicate (CSH) and calcium aluminate (CAH) besides calcium hydroxide (CH). These hydrates are generally referred as tobermorite gel. Adhesive and cohesive properties of the gel bind the aggregate particles. With time the cement paste exhibits setting and hardening which impart concrete its properties in fresh and hardened state.

Calcium hydroxide is really a by-product of cement hydration (Neville 1981). When fly ash in incorporated in concrete, the calcium hydroxide liberated during hydration of Portland cement reacts slowly with the amorphous alumino-silicates, the pozzolanic compounds, present in the fly ash. The products of these reactions, termed as pozzolanic reaction products, are time dependent but are basically of the same type and characteristics as the products of the cement hydration. Thus additional cementitious products become available which impart additional strength to concrete. Because the pozzolanic reactions are much slower than the cement hydration reactions, partial replacement of the cement in concrete generally reduces early strength, but may be equal or increase the long term strength. The rate of strength gain, however, depends upon properties of the fly ash and cement used, mix proportions, as well as the curing conditions of the fly ash concrete (Joshi 1979).

Concrete mixes are designed with more water than needed for cement hydration to obtain proper workability. This excess water is present in capillary channels of the hydrated cement paste and is commonly referred as capillary water. In a properly cured concrete, the calcium hydroxide dissolved in the capillary water would react with the fly ash to form the solid reaction products that will fill partially or completely the capillary channels. The blocking of the capillary pore/channels both by physical action due to fine particle size and due to the formation of new products of pozzolanic reaction results in lower permeability of concrete. The reduced permeability also reduces the aggressive and deleterious action of the salt solutions such as chloride or sulphate solutions.

Another important aspect is that the cement hydration reactions are exothermic in that a portion of the latent energy required to combine the elements is released by the hydration reaction. As a result the temperature of concrete is raised. Initial setting or hydration of cement rapidly increase the temperature within a large mass of concrete since the heat is not dissipated quickly enough. The subsequent cooling of the concrete introduces internal stresses of sufficient magnitude to cause cracking. With partial replacement of cement by fly ash in concrete the heat of hydration is reduced. Further the heat is released over a long period of time because of the reduced amount of cement and slower pozzolanic reactions. Thus the temperatures in mass concrete, in particular, remain lower because heat is dissipated as it develops.

As a result of extensive laboratory and field research, the use of fly ash as pozzolan in cement concrete has been well established and recognized by concrete industry. When the fly ash also has cementitious properties, as is the case with Class C fly ashes, additional strength producing reactions also occur. These reactions although complex but are generally considered similar to normal hydration reactions of Portland cement. Therefore, the heat of hydration may not reduce significantly when high calcium Class C fly ash is used for partial replacement of cement in concrete.

3.2. METHODS OF CONCRETE MIX PROPORTIONING WITH FLY ASH

In a concrete mix, fly ash can act in part as a fine aggregate and in part as a cementitious component due to its pozzolanic or cementitious properties. The earlier studies (Smith 1967), related to the proportioning of fly ash concrete, were based on the assumption that every fly ash possesses a unique cementing efficiency K, such that when multiplied by the mass of fly ash would be equivalent to the mass of cement. Thus cementitious material in a given concrete is later equal to the cement and KF, where F is the mass of fly ash. However, later studies have indicated that K shows wide variations and thus is not suitable as a design parameter. The value of K also varies with the type and amount of cement in addition to curing conditions and the strength level of the concrete. Different water demands for different fly ashes also require adjustments of aggregate contents of concrete. This approach was found complex and impractical for most applications.

Lovewell and Washa (1958) and several other investigators have developed methods to proportion fly ash concrete to have strength at any desired age equal to that of comparable conventional concretes. The underlying principle originally used by Lovewell and Washa was "in order to obtain approximately equal compressive strength between 3 and 28 days, mixes made with fly ash must have a total weight of Portland cement and fly ash greater than the weight of the cement used in the comparable strength Portland cement mixes."

A critical review of various procedures for incorporating fly ash into concrete indicates that in practice, fly ash can be added either as an admixture at the concrete mixer or as an ingredient in blended cement. In the latter case, the ratio of fly ash to Portland cement remains fixed and generally no adjustment in amount of cementitious materials is made when blended cement is used in place of ordinary Portland cement. Addition of fly ash at the mixer provides opportunities for adjustment of the ratio of fly ash to cement and, therefore, makes it possible to use optimum proportions of specific fly ash with different type of cements. Whether added as a portion of the blended cement or at the ready mix plant, the effect of a particular fly ash with the same cement should be essentially the same for the same rates and amounts of cementitious materials. In North America, the present trend is towards adding fly ash as an admixture at the concrete mixer in the form of a separately batched concrete ingredient.

In proportioning fly ash concrete mixes reference is generally made to an equivalent plain concrete mix for defining the requirements for mix design. Generally the following mix proportioning methods have been advanced.

3.2.1. Simple Replacement Method

This method is based on partial replacement of cement where certain mass or volume of fly ash replaces an equal mass or volume of cement. The normal result of this procedure reported in the literature is a concrete that has lower strength at early age, possibly up to 90 days, but thereafter usually has higher strength than the control concrete made with Portland cement. This is in agreement with the generally believed view that at early ages fly ash exhibits very little cementing action and acts rather as a fine aggregate. But at later ages pozzolanic activity becomes apparent and contributes considerably to strength gain in concrete containing fly ash. In recent studies on Class C fly ashes, the above belief has been refuted. Use of Class C fly ash does not always reduce early strength. Prior to 1975, mostly low calcium Class F fly ashes, were used for replacement of cement. As a result, the early strength was low and ultimate strength equal to or higher than normal Portland cement concrete.

In mass concrete constructions where fly ash was initially used early age strength was of little concern in the light of the desired reduction in temperature rise due to heat of hydration. The replacement methods of mix proportioning were, therefore, commonly used. These methods result in cost saving as well as the reduction of the heat of hydration almost in direct proportion to the amount of cement replaced with fly ash and the ratio of cost of fly ash to that of cement at the construction site. With the advancement and use of chemical admixtures as well as production and availability of Class C fly ashes, it has been possible to develop high strength and durable concrete mixes by using simple replacement of cement with fly ash in the range of 50 to 60% by weight.

3.2.2. Addition Method

In this method, fly ash is added to the concrete as fine aggregate without a corresponding reduction in the quantity of cement used. In the construction of the South Saskatchewan River Dam in Canada (1953) this approach was used to improve the sulphate resistance of concrete. With long periods of moist curing, the effective cementitious content of the concrete is increased by this method. Because of the pozzolanic activity of fly ash, significant improvement in strength beyond early ages of 7 days is generally obtained. This method does not use the cementing potential of fly ash and results in an uneconomical fly ash concrete mix. However, for imparting specific properties to concrete for a particular job, the addition of fly ash as partial replacement of fine aggregate may be adopted in mix proportioning with suitable adjustments in aggregate content as well as water content.

3.2.3. Modified Replacement Methods

Most of the current standards and guide lines for proportioning of fly ash concrete are based on the procedures involving the concept originally advanced by Lovewell and Washa and using Abrams relationship between strength and water to cementitious material ratio (Abrams 1918, Lovewell and Washa 1958). Over the years research on fly ash characteristics and their effects on properties of cement concrete have led to the development of improved proportioning procedures for fly ash concrete mixes for various purposes. For practical applications, the fly ash concrete mix proportioning is controlled by a number of factors. Some important factors and underlying assumptions are given below.

Strength development in concrete using a given fly ash may vary with different cements just as strength development of different fly ashes will be different with the same cement. The earlier concept that fly ash addition usually reduced the strength of concrete at early age may not be relevant to currently available high quality fly ashes, specially Class C fly ashes, produced in modern thermal power plants burning high calcium coal and using special equipment and procedures for coal combustion and dust collection.

For equal workability, Class F fly ash addition usually reduces the water demand that is required for plain cement concrete. On the other hand use of Class C fly ash may not affect the water demand and may even increase it sometimes. However, fly ash concrete is more sensitive to curing conditions than plain concrete and needs relatively longer curing.

As in normal concrete mix proportioning, trial mixes must be made and tested with fly ash concrete. Such trial mixes should include the chemical admixtures to be used as well as the cement, fly ash and aggregate. Data on each fly ash must be developed as an aid to trial mix proportioning. Results of trial mixes should ensure that the design characteristics of fly ash concrete are attained.

In assessing the relative economy of fly ash mix, allowance should be made for the handling of an extra component when fly ash is used as an admixture and batched separately. At present, two approaches are suggested for proportioning fly ash concrete mixes. One is to modify a control mix so that the concrete containing fly ash as partial replacement of cement will have corresponding properties of workability and strength, the assumption being that other properties of the concrete will be comparable or better. The ultimate goal in this approach is to proportion fly ash concrete with reference to a control mix. It is of course recognized that more research is needed to better understand the cement — fly ash interactions before this goal can be reached.

The Electric Power Research Institute in the United States and CANMET in Canada besides several other organizations are conducting research to relate concrete performance to the physical and chemical characteristics of the fly ashes and to develop predictive models. The current studies should ultimately provide better proportioning procedures to achieve maximum economy and optimum properties of fly ash concrete.

3.3. PERFORMANCE TESTS OF FLY ASH CONCRETE

Generally the use of fly ash in concrete for partial replacement of cement is as a pozzolan. The maximum amount of cement that can be replaced with fly ash is normally specified by the construction agency depending on the quality and characteristics of fly ash from a given source as well as the purpose for which concrete is designed. Usually the mass of fly ash used is relatively more than the cement replaced with additional adjustments in water and fine aggregate to obtain optimum characteristics of concrete.

Review of the literature reveals that the maximum replacement levels of cement with fly ash vary from 8 to 60%, with 30 to 60% levels commonly adopted for mass concrete to reduce the heat of hydration. For field concrete applications, specifications are commonly based on the minimum cement or cementitious material (cement and fly ash) content as well as on the specified water-to-cementitious material ratio. For bridge deck concrete ACI committee 345 specifies minimum cementitious material content of 335 kg/m^3 and water-to-cementitious material ratio no greater than 0.45.

Essentially quality control tests and procedures used for acceptance of fly ash concrete are the same as employed for the equivalent plain Portland cement concrete without fly ash. In fresh state, the main tests conducted are setting time (ASTM C403), slump/workability (ASTM C143), air content (ASTM C231), and unit weight (ASTM C138). Whereas for the hardened concrete, the tests include comprehensive strength (ASTM C39), flexural strength (ASTM C78), modulus of elasticity (ASTM C469), shrinkage (ASTM C157), and freeze-thaw durability (ASTM C666). Additionally, several other control tests for specific job requirements may also be specified. For wet or fresh concrete, bleeding and segregation and temperature test; and for hardened concrete tests for creep, resistance to chloride and sulphate solution attack, scaling resistance against deicers, resistance to alkali aggregate reactions, fire resistance, resistance to abrasion and erosion of concrete, etc. may also be required.

Ideally, a performance specification based on the strength and durability of the concrete should be used. However, there are uncertainties about the ability of present tests, such as the rate of strength development, resistance to freezing and thawing because of the erratic amounts of entrained air content as well as soundness, to predict the strength and overall durability of concrete containing fly ash as a replacement for part of the usually specified amounts of cement. The role of chemical ad-mixtures may also show varying effects in concretes with and without fly ash.

Quality assurance procedures for fly ash concrete include quality control of the fly ash along with the selection of other concrete ingredients, rational mix proportioning and placement practices. Of equal importance is economics. The distance that fly ash must be transported to reach the point of use has a significant effect on whether the use of fly ash in concrete is economically feasible. The potential volume of fly ash concrete for a particular project would also affect the cost effectiveness because of the capital investment involved in silos and fly ash handling equipment.

The American Coal Ash Association has issued guidelines for a fly ash quality assurance program. Similarly, the Electric Power Research Institute has completed a project for fly ash classification to obtain the full spectrum of characteristics of North American fly ashes. The present special equipment and efforts to produce quality fly ash by power companies and the adoption of judicious quality assurance programs for fly ash as well as fly ash concrete by the user/construction agency should assure that the resultant concrete is satisfactory. Regular quality assurance tests such as uniformity tests are performed to ensure that ash of specified density, fineness, and loss of ignition, from a given source is suitable as established by previous tests on trial concrete mixes made with the given fly ash. However, the ash from a new source must be extensively tested through trial mixes for satisfactory performance in concrete. Once the source of fly ash is approved, the testing need not be performed on every delivered batch of ash. However, to ensure suitability of fly ash concrete for any given field application, some routine testing involving determination of air content and rate of strength development is generally specified at the batch level. The present state of knowledge requires that until better understanding is available concerning the load-to-load uniformity of air content in the fly ash concrete, test for air content should be made on each load of concrete immediately before its placement.

3.3.1. Strength Requirement

Usually all specifications for fly ash concrete for field use include 28 day strength requirements for concrete, the same as those for Portland cement concrete. The reliance of accelerated strength test at earlier ages for acceptance must be determined by experience. Any concrete that meets present criteria for acceptable strength at early ages would be acceptable but because of the potentially slower strength development, many acceptable fly ash concretes may not reach the strength level established by experience for Portland cement concrete. Sometimes, strength tests at 90 days may be specified to establish the level of strength for mature fly ash concrete. Currently for many purposes specifications permit the use of water reducers or superplasticizers to lower the ratio of water to cementitious material and achieve design strength in the fly ash concrete, at early ages of 7 to 14 days, equal to the corresponding strength of concretes without fly ash but with higher corresponding water to cement ratio.

3.3.2. Air Entrainment in Fly Ash Concrete

The commonly experienced difficulty of ensuring adequate entrained air in the hardened fly ash concrete is a major concern for all construction agencies using the concrete in cold climatic regions. To improve freeze–thaw durability of concrete air in the form of very closely spaced minute bubbles is introduced in the whole mass of concrete by using air entraining admixture. Erratic results with respect to air entrainment are observed in many cases while using fly ash in concrete. Inadequate air entrainment results in poor performance of fly ash concrete under freezing and thawing conditions. Slight variations in the quality of fly ash, especially carbon content or LOI exhibit wide variations in the amount of entrained air content for the same ingredients. This has led to a reluctance to continued and increased use of fly ash. Until more knowledge is gained related to the rate of loss of air content in the fly ash concrete and suitable measures are developed to ensure adequate air content at the time of placement increased use of fly ash in concrete is likely to be hampered. It is reported by many investigators that the unburned carbon in fly ash is possibly the source of variability of air entraining agent demand. On the basis of present knowledge, a completely safe limit on LOI or unburned carbon cannot be established because the erratic behaviour of air entrainment in concrete has been observed even with Class C fly ashes with LOI or carbon content less than 1.0%.

Different cements and different air entraining agents could react differently with the same proportions of other ingredients in the fly ash concrete. For quality control of fly ash concrete, tests for air content is required on each batch of concrete immediately before its placement. For quick detection of significant variations in air content, the average of two results obtained by a Chace air indicator, carefully calibrated against the air pressure meter, provides adequate control. For any batch of concrete for which compliance to the specification for air content is in doubt when determined by the Chace indicator, a determination needs to be made by the air pressure meter and the decision to accept or reject the concrete made on the basis of that test. It is extremely important to test each load of concrete for air content until the basic reasons for the variations of entrained air in the fly ash concrete are understood. The durability of the concrete is at stake and the replacement costs of the failed concrete can be prohibitively high.

CHAPTER 4

Effects of Fly Ash on the Properties of Fresh Concrete

4.1. GENERAL

The present state of knowledge recognizes the usage of fly ash in cement and concrete as raw material in cement production, as an ingredient in blended cement, and as mineral admixture in concrete. Sometimes fly ash is also used as partial replacement of fines. Based on laboratory investigations as well as field applications of fly ash concrete over the last 50 years, several comprehensive reviews and other publications present the accepted views related to the advantages and disadvantages of incorporating fly ash in concrete.

The majority of the recognized effects of fly ash on concrete properties tends to improve concrete performance in field use. The addition of fly ash to concrete affects its properties both in the fresh and hardened states favourably. The nature and degree of effect on a specific concrete property, however, depends upon several factors such as type and amount of fly ash, mix proportion, chemical admixtures, curing conditions, and other job requirements including construction practices.

In fresh concrete, fly ash plays an important role in the fluidity of concrete which is commonly expressed in such phenomenological measurements as workability, pumpability, compactability, water demand, bleeding and segregation, and finish-ability. Addition of fly ash has significant influence on the rate of hydration reactions as well as on the effectiveness of the chemical admixtures, particularly air entraining agent, and water reducer or superplasticizer.

Low calcium Class F fly ash normally acts as a fine aggre-gate of spherical form in early stages of hydration whereas high calcium Class C fly ash may contribute to the early cementing reactions in addition to its presence as fine particulate in the concrete mix. Hydration of cement is an exothermic reaction and the released heat causes a rise of temperature of fresh concrete. For producing high strength concrete, high range water reducer or superplasticizer is added to maintain the given workability of concrete at a low water–cement ratio. Further-more, air entrainment of adequate amount, usually $6 \pm 1\%$, is obtained by using air entraining admixture to improve freeze–thaw durability of concrete. As fly ash forms one of the com-ponents of concrete, its effects on the general properties of fresh concrete need better understanding. In this section, therefore, an account of the role of fly ash on the rheology, temperature rise, water demand, setting time, and air entrainment, of fresh con-crete is presented.

4.2. WATER REQUIREMENT AND WORKABILITY

Workability is generally one of the governing requirements of concrete mix proportioning. The unique definition of this prop-erty has defied the efforts of several investigators. In most cases, the fluidity of concrete is termed as workability and is defined as the ease with which concrete can be mixed, placed, handled, compacted and finished. Workability is determined by the rheo-logical behaviour of fresh concrete and in fact is the cardinal property of fresh concrete. Water content of concrete plays a dominant role in controlling workability. Therefore all those factors which affect water requirement in concrete will neces-sarily influence its workability.

The spherical shape and glassy surface of most fly ash particles, usually finer than cement, permit greater workability or slump for equal water–cement ratios. In other wards, the water cement ratio can be reduced for equal workability in a concrete mix by including fly ash. In practice, because the absolute volume of cement plus fly ash, especially Class F ash, normally exceeds that of cement in similar concrete mixes without fly ash, the increased ratio of solids volume to the water volume produces a paste with improved rheological properties such as plasticity and the cohesiveness. This results in the stability of the dispersion of the cement and fly ash particles in the fresh paste which act as binder in the concrete mass.

Several investigations have been conducted to study the role of Class F fly ash on the rheology of fresh concrete (Hobbs 1983a). In two field applications (Compton and McInnis 1952) 30% fly ash substitution for cement was found to reduce the water demand by 7% at constant slump. Brown (1952) conducted several studies with fly ash replacing cement and/fine aggregate at levels of 10 to 40% by volume. According to Brown (1952) each 10% of ash substituted for cement, the compacting factor or workability changed to the same order as it would by increasing the water content of the mix by 3 to 4%. When fly ash was substituted for sand or total aggregate, workability increased to reach a maximum value at about 8% ash by volume of aggregate. Further substitution caused rapid decrease in workability.

Price (1953) reported that water demand was not increased with the addition of fly ash to the concrete made at fixed cement contents for the construction of the South Saskatchewan River Dam in Canada. Locally available lignite fly ash was used as a replacement of fine aggregate rather than cement in concrete at this dam site. The resulting concrete at a given water content had a lower ratio of water to total cementitious material, yet workability and cohesiveness of the mix were improved.

Carette and Malhotra (1984) studied the effects of a number of Canadian fly ashes on the general properties of fresh concrete. The results of their study are presented in Table 4.1. The air entrained concrete mixes were prepared using different fly ashes at replacement level of 20% by mass of cement. The mixes were proportioned to contain equal ratios of water to total cementitious material, containing cement and fly ash, which was kept constant. From the results in Table 4.1 it is observed that slump does not always increase with the incorporation of fly ash in concrete. In general, the fly ash addition exhibited improvement in slump in the tested concrete mixes as compared with that of control mix.

Welsh and Burton (1958) reported loss of slump and flow for concretes made with some Australian fly ashes when used as partial replacement of cement at a constant water content. Similarly the experience with Indian fly ashes and Canadian fly ashes reported by Lohtia et al. (1973, 1974) and Joshi (1973), respectively, suggested that in some cases fly ash addition to concrete caused an increase in water demand to maintain constant slump. All the fly ashes may thus not always reduce water demand in concrete. Generally, the fly ashes from older power plants with high carbon content and coarse particle size show increased water requirement to maintain constant workability. Owens (1979) and many other researchers have reported that with the use of fly ash, containing large fraction of particles coarser than 45 µm or a fly ash with high amount of unburned carbon, exhibiting loss on ignition more than 1%, increasing water demand is observed. Water demand is noticeably increased to maintain the desired level of fluidity. Figure 4.1 illustrates the effects of coarse fly ash particles on water demand.

Brink and Halstead (1956) reported that some fly ashes reduced the water demand of test mortars, while others especially those containing higher carbon content showed increased water requirement compared to those of control mortars. In a study at CANMET (1984) of the eleven fly ashes from modern Canadian power plants, nine showed significant improvement in workability at constant water content.

TABLE 4.1 Mix proportions and properties of concretes incorporating some Canadian fly ashes (Carette and Malhotra 1984)

Mix no.	Cement (kg/m³)	Fly ash (kg/m³)	Fine agg. (kg/m³)	Coarse agg. (kg/m³)	AEA (ml/m³)	W/(C+F)	Slump (mm)	Air (%)	Unit weight (kg/m³)	Bleeding (%)	Setting time (h:mm) Initial	Final
Control	295	0	782	1082	170	0.50	70	6.4	2320	2.9	4:10	6:00
F1	236	59	780	1077	320	0.50	100	6.2	2300	3.1	4:50	8:00
F2	237	59	782	1080	200	0.50	105	6.2	2310	4.6	7:15	10:15
F3	238	59	786	1088	200	0.50	100	6.2	2310	5.1	5:20	8:10
F4	237	59	792	1094	160	0.50	110	6.3	2320	4.3	6:20	8:25
F5	238	59	782	1080	690	0.50	65	6.4	2310	2.7	5:15	8:55
F6	238	59	784	1082	660	0.50	75	6.5	2300	2.6	4:30	6:50
F7	239	59	780	1077	370	0.50	100	6.1	2300	2.9	4:15	6:20
F8	236	59	775	1069	230	0.50	115	6.2	2300	5.6	5:10	7:30
F9	236	59	775	1070	240	0.50	100	6.4	2280	4.4	5:25	9:00
F10	237	59	781	1079	290	0.50	130	6.5	2290	2.5	4:45	7:00
F11	237	59	782	1080	150	0.50	140	6.6	2290	0.6	4:00	6:05

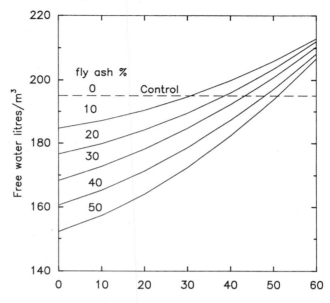

Fly ash particles greater than 45um, % by mass

FIGURE 4.1   Influence of coarse particulate content of fly ash on the
             water required for equal workability in concrete
             (Owens 1979).

In an extensive study to produce high strength and durable
concrete mixes with high replacement levels of three bituminous fly ashes from Alberta, Joshi et al. (1990, 1992) conducted
tests on 70 trial concrete mixes in which chemical additives
such as superplasticizer, and air entraining agent were also
included. Table 4.2 presents the mix proportions used and the
properties of the concrete mixes in fresh state. The data in Table
4.2 show increased workability of concrete mixes containing fly
ashes compared to the control mixes without fly ash for the
same quantity of superplasticizer and at the fixed ratio of water
to cementitious material. Because of the relative coarse nature
of fly ash particle sizes, Wabamun fly ash concrete mixes had

TABLE 4.2 Properties of fresh concrete mixes (Joshi et al. 1987)

Chemical admixture type	Mix no.	Mineral admixture type	Slump (mm)	VEBE (s)	Density (kg/m³)	Air (%)	AEA ml per kg cementitious material
	1A	Control	5	6.9	2422	2.8	
	2A	Forestburg	20	5.4	2370	2.2	
	2A	Sundance	10	5.8	2387	1.8	
	4A	Wabamun	0	6.3	2387	1.7	
No	5A	Boundary Dam	20	5.4	2401	1.7	
Chemical	6A	Laramie	20	5.5	2402	2.1	
Admixture	7A	Lakeview	0	5.5	2387	2.2	
	8A	Nanticoke	0	5.6	2394	2.7	
	9A	Lingan	25	4.8	2401	2.0	
	10A	limestone	5	6.6	2401	1.8	
	11A	Silica	0	10.6	2415	2.1	
	1B	Control	25	4.6	2415	1.8	
	2B	Forestburg	120	2.2	2387	1.9	
	3B	Sundance	170	1.4	2429	0.6	
	4B	Wabamun	65	1.8	2401	1.7	
Water	5B	Boundary Dam	125	1.2	2429	1.3	
Reducer	6B	Laramie	110	1.9	2387	2.2	
Only	7B	Lakeview	0	4.6	2430	2.5	
	8B	Nanticoke	20	4.6	2401	2.5	
	9B	limestone	35	3.4	2394	2.5	
	10B	Silica	30	3.1	2430	1.4	
	1C	Control	20	6.7	2274	7.1	0.7
	2C	Forestburg	25	4.0	2288	6.0	1.3
	3C	Sundance	50	2.0	2267	6.6	2.2
Air	4C	Wabamun	30	2.1	2281	6.0	1.6
Entraining	5C	Boundary Dam	80	1.3	2274	6.2	3.8
Agent	6C	Laramie	50	2.8	2295	7.0	1.0
Only	7C	Lakeview	0	4.7	2302	6.4	8.0
	8C	Nanticoke	15	3.7	2323	6.0	5.4
	9C	Lingan	60	1.3	2295	6.6	0.9
	10C	limestone	30	2.3	2288	6.5	1.8
	11C	Silica	18	5.6	2260	6.5	2.1
	1D	Control	60	5.2	2316	6.0	0.7
	2D	Forestburg	110	1.6	2196	7.4	1.0
Water	3D	Sundance	120	1.6	2316	6.5	1.4
Reducer	4D	Wabamun	120	0.9	2260	6.1	1.2
and	5D	Boundary Dam	150	0.8	2302	6.2	2.2
Air	6D	Laramie	165	1.5	2401	7.2	1.0
Entraining	7D	Lakeview	5	4.3	2302	6.0	5.2
Agent	8D	Nanticoke	20	3.2	2302	6.0	2.0
	9D	Lingan	175	0.4	2288	6.1	1.3
	10D	limestone	55	1.9	2290	5.8	1.5
	11D	Silica	40	2.0	2302	6.6	1.7

50% replacement of cement by mineral admixture in all mixes except control mixes 1A, 1B 1C and 1D. In mixes 2–9 except 9B, various type of fly ashes used. For all the mixes: water to cementitious material ratio = 0.47; water reducer used = 7.11 ml per kg of cementitious material.

lower slump than the corresponding mixes made with Sundance and Forestburg fly ashes. But Wabamun fly ash mixes still showed better workability than the equivalent plain concrete mix. In general, the use of fly ash as replacement of cement produced better workability with a marked improvement in the cohesiveness and finishability of the concrete mix.

To develop a theoretical understanding of the rheology of fresh fly ash concrete, Tattersall (1976), Tattersall and Benfill (1983) and other researchers have suggested mathematical expressions relating the yield value (τ) and the plastic viscosity (μ) to volumetric parameters of concrete in terms of a Bingham model. It is hypothesized that the yield value and the plastic viscosity decrease as the volume of fly ash increases until minima are reached and then increase as the aggregate replacement level is further increased. The research studies related to the effect of fly ash on workability of concrete reveal the applicability of the Bingham model to explain the observed behaviour. Ivanhov (1982) reported that "Yield stress (τ) and plastic viscosity (μ) varied with volumetric parameters of concrete, water to cement ratio, replacement levels of cement/aggregates by fly ash, and fineness of fly ash. An increase in the volume of the paste at a constant ratio of ash to total cementitious material resulted in an increase in plastic viscosity." Apparently because of fine particle size and smooth glassy texture as well as spherical shape, fly ash acts to plasticize concrete at a given water content when used as partial replacement of cement or fine aggregate.

4.3. SEGREGATION AND BLEEDING

In general, bleeding and segregation are considerably reduced with the use of fly ash as a mineral admixture in concrete and thus pumpability of concrete is improved. The lubricating effect of the glassy spherical fly ash particles and the increased ratio of solids to liquid make the concrete less prone to segregation and increase concrete pumpability. Figure 4.2 shows the bleed-

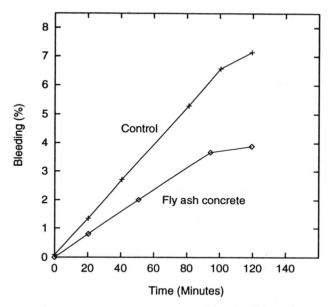

FIGURE 4.2   Relative bleeding of control and fly ash concretes
(CGEB 1967).

ing rate of fly ash concrete compared to that of control concrete
without fly ash.

The use of fly ash particularly in the harsh mixes, which are
deficient in fines, would resolve the problem of excessive bleeding by increasing the overall paste volume by the addition of
fly ash in concrete as mineral admixture (Johnston 1994).
Carette and Malhotra (1984) reported that six of the eleven
Canadian fly ashes had increased bleeding as compared to that
of control concrete as shown in Table 4.1.

From a recent study on high volume fly ash concrete mixes,
using Alberta fly ashes, Joshi and Lohtia (1993a) reported that
the fly ash concrete mixes were more cohesive than control
mixes. During the slump test, the fly ash concrete mixes sub-

sided more slowly and gradually than the control mixes which exhibited abrupt fall or subsidence.

When fly ash is used in concrete, workability can be increased to the point that sand content can be decreased and coarse aggregate content increased, thereby reducing the total surface area to be coated with the binder matrix. This leads to further improvement in workability for the given amount of cementitious material. Because of the increased fine particulate content, some fly ashes provide a marked improvement in concrete finish when used as partial replacement of cement or fine aggregate. In general, fly ash is found to be particularly useful and valuable when used in lean mixes and in concretes made with aggregates deficient in fines, as it improves their workability and finishability. A more workable concrete can be compacted to a higher degree with the given amount of effort that would result in increased strength and enhanced durability of the hardened concrete.

4.4. TIME OF SETTING

Immediately after adding water to concrete, hydration and other chemical reactions start and the cement paste begins to stiffen accompanied by heat release. The rate of stiffening of cement paste is expressed in terms of setting time. The terms initial and final set are referred to arbitrarily chosen stages of setting and are determined by some form of penetrometer test (ASTM 403-92).

The majority of the investigations reveal that the addition of low calcium Class F fly ashes generally shows some degree of retarding effect on cement setting. Lane and Best (1982) reported the retarding influence of fly ash on setting of concrete as function of the proportion, fineness and chemical composition of the fly ash. They concluded that the influence of fly ash on setting time is less than the influence due to cement fineness, the water content of the paste, and the ambient temperature.

From the date in Table 4.1, it is seen that all except one of the eleven Canadian fly ashes studied markedly increased both the initial and final setting times. The fly ashes used in the study had CaO content varying from 1.4 to 13.0%. At low temperature, cement setting is retarded both in plain and fly ash concrete. A study by Mailvaganam, Bhagtath and Shaw (1983) on fly ash concrete properties at two different temperatures showed that the concrete mixed at 5°C exhibited retardation in excess of 10 hours, regardless of fly ash content. The test concretes in their study were made with low calcium Class F fly ash and using various admixtures. At 20°C, the concretes containing 30% fly ash by weight of cement showed setting times extended by about 1 to 1.75 hours.

Normally, the effect of fly ash on the time of setting depends on the characteristics and amount of fly ash used. The interacting effects of fly ash with other chemical and mineral admixtures used in concrete may also influence the setting of concrete. All Class F fly ashes, especially low calcium fly ashes with high carbon content or LOI, increase the setting time.

The high calcium fly ashes which are generally low in carbon and high in reactive and/or cementitious components sometimes exhibit opposite behaviour of reduced setting time. Not all Class C fly ashes cause rapid setting. Ramakrishan et al. (1981) and Rodway and Fedirko (1989) reported an increase in setting time with the use of high calcium fly ash in concrete.

Table 4.3 presents test data of the fly ash concrete mixes tested by Rodway and Fedirko (1989). The concretes were made with Alberta fly ash of amounts varying from 0 to 76% of the total cementitious material and using superplasticizer and air entraining admixture to maintain constant slump and air content within specified limits. High fly ash concrete mixes exhibited increasingly greater initial setting times of 22 to 42.5 hours with increasing fly ash content from 56 to 76% compared to 7.6 hours for the control mix without fly ash. The observed delays appeared to be related to the problem of compatibility between cementitious materials and superplasticizer to maintain

TABLE 4.3   Mix proportions, properties of fresh concrete and summary of compressive strength of Western Canadian fly ash concrete (Rodway and Fedirko 1985)

Mix no.	1	2	3	4
W/C+F	0.41	0.28	0.28	0.28
F/C+F	0	0.56	0.68	0.76
Batch quantities kg/m^3				
Cement (ASTM Type 1)	275	145	110	82
Fly Ash (ASTM Class F)	0	185	235	258
Sand	845	800	800	775
Gravel	1100	1100	1100	1100
Water	112	93	97	95
Air entraining agent (ml/m^3)	470	250	375	250
Water reducer (ml/m^3)	1290	690	517	402
Superplasticizer (ml/m^3)	5.2	8.0	9.0	9.0
Properties of fresh concrete				
Temperature (°C)	14	14.5	14	14
Slump (mm)	140	125	150	150
Air content (%)	5.0	5.8	5.6	6.0
Density (kg/m^3)[a]	2362	2374	2339	2317
Initial setting time (h:min)	7:32	22:09	30:45	42:44
Compressive strength (MPa)				
152 × 305 mm test cylinders[b]				
1 day	9.8	–	–	–
3 days	25.4	18.5	6.4	2.4
7 days	31.6	28.9	17.6	7.7
28 days	36.7	50.6	39.8	22.7
56 days	40.5	59.3	49.2	28.4
91 days	42.1	65.8	56.3	31.3
102 × 203 mm drilled cores				
7 days	27.2	24.8	15.3	8.5
28 days	34.0	36.1	27.3	19.4
56 days	33.3	39.0	34.5	23.6
91 days	37.4	43.7	39.2	26.6

[a]At the time of demolding. [b]Each value is average of two cylinders.

workability. The observed significant set retardation was attributed to higher dosages of superplasticizers used. The same investigators later modified the mix proportions to regulate and control the amount of plasticizer for field use. The resulting concrete mixes had initial time of set around 7.5 hours comparable to that of the control concrete (Rodway and Fedirko 1992).

The test results of high volume fly ash concrete mixes studied by Sivasundaram et al. (1989b, 1990) showed that the initial time of set of 7.50 hours is comparable to that of the control concrete, whereas the final setting time is extended by about 3 hours as compared to that of the control.

In a recent study (Joshi et al. 1993) involving three different sub-bituminous Alberta coal ashes used at replacement levels of 40 to 60% by cement weight to produce superplasticized and air entrained concretes, it was observed that fly ash concrete achieved an initial set in 5 to 11 hours as compared to about 5 hours for non fly ash concrete. The final setting time varied from 10 to 13 hours as against 7 hours for control mixes containing no fly ash.

The observed set retardation with fly ash addition is, in general, not of much significance particularly for mass concrete construction and highway construction. However, for slip forming as well as cold weather concreting, setting time needs to be accelerated when fly ash concrete is involved. The delayed set can be of benefit during summer months specially in hot weather concreting.

4.5. AIR ENTRAINMENT

Minute air bubbles are intentionally entrained in concrete by using air entraining agent (AEA) to improve the freeze–thaw durability of concrete. For entraining a specified amount of air content, usually around 6%, more air entraining agent is required in fly ash concrete than for a similar concrete containing no fly ash. This increase is reported to be because of the greater surface area of the fly ash in concrete. Fly ash is generally finer than cement and the volume of fly ash added is normally more than the volume of the cement replaced. Due to this, a greater volume of air entraining agent is needed to provide the same surface concentrations of the air entraining agent. The second, and possibly the main, reason leading to the increased demand of AEA is related to the carbon content, normally expressed in terms of loss on ignition of fly ash. Since 1980, loss on ignition limit in ASTM C618 has been reduced to 6.0% from the earlier permitted limit of 12% for the effective use of fly ash in air entrained concrete.

Recent studies on Alberta fly ashes have indicated that the problem of erratic air entrainment are encountered even with the ashes with carbon content less than 0.5% (Joshi et al. 1991). Thus not only the amount but the form in which carbon is present in fly ash possibly also affects the AEA demand in fly ash concrete. The carbon absorbs a portion of the AEA, and thus makes it unavailable for creating the needed conditions for stable air bubbles. The presence of adsorptive or activated carbon with high specific surface may also alter the effectiveness of other admixtures. Significant loss of air with time, erratic air entrainment and especially the increased AEA demand may sometime offset the possible benefits with the utilization of fly ash in concrete. Because of this problem, at times many concrete producers and construction agencies have shied away from the fly ash concrete or ash use in concrete (Joshi and Lohtia 1995). Loss on ignition or carbon content is still considered as the influencing factor for variability of AEA demand in fly ash concrete. Some construction agencies have adopted a lower limit for loss on ignition (LOI), usually less than 3.0% to avoid the problem. The AASHTO specification M295 places the limit at 5.0% as against 6.0% in the relevant ASTM standards.

It has been reported that concretes made with Class C fly ash generally require less AEA than those made with Class F fly ashes (Gebler and Klieger 1983, 1986a,b). For some Class F fly ashes replacing about 50% cement, the average requirement of AEA is found to be more than double that of the equivalent plain concrete (Joshi et al. 1993). Furthermore, Alberta Class F fly ashes (Hague et al. 1988a) seem to show different patterns of behaviour on air entrainment than that of the Class C fly ashes studied by Gebler and Klieger (1983).

The amount of AEA used in a recent study on superplasticized and air entrained concrete mixes made with different amounts of Alberta fly ash in the range of 40 to 60% are given in Table 4.4 (Joshi and Akkad-Salam 1992). In this study, air content of the freshly mixed concrete was determined according to the procedure described in ASTM C231 by the "pressure method" using type B air meter. Neutralized vinsol resin (Daravair) meeting the requirements of ASTM C260 was used as air entraining admixture to entrain around 6% air content in concrete.

A wide range of AEA demand was reported by Carette and Malhotra (1984) from the study of concretes made with Canadian fly ashes to entrain about 6% air content. Gebler and Klieger (1983) reported that for 6% air content in concrete, the AEA varied from 126 to 173% for fly ashes having more than 10% CaO, whereas it was in the range of 177 to 553% for fly ashes containing less than 10% CaO.

Air content, loss of air with time and air void systems parameters in the hardened concrete are the primary considerations in air entrained concrete with respect to its frost resistance. Fly ash addition affects both air content and loss of air content with time in plastic or fresh concrete depending upon the type of fly ash. Gebler and Klieger further suggest that increase in both total alkalies and SO_3 contents in fly ash affect the air entrainment favourably. A concrete containing a Class F fly ash that has relative high CaO content and less organic matter or carbon tends to be less vulnerable to loss of air.

Not only fly ash, but also other mineral admixtures such as blast furnace slag, silica fume, rice husk ash etc. cause increase in AEA demand possibly because they are much finer than Portland cement. However, the variability due to fly ash addi-

TABLE 4.4 Proportions and properties of final design mixes using high volume fly ash (Akkad Salam 1992)

Mix no.	Fly ash[a] (kg/m³)	Cement[b] (kg/m³)	Coarse agg. (kg/m³)	Fine agg. (kg/m³)	$\frac{W}{C+F}$ [c]	AEA[d] (ml)	Super-plasticizer[e] (ml)	Density (kg/m³)	Slump (mm)	Air content (%)	Compressive strength (MPa) 7 days	28 days
FS40-I	186(S)	280(I)	1088	700	0.28	130	1125	2258	120	5.2	39.8	47.0
FS50-I	233(S)	233(I)	1088	700	0.28	130	900	2206	140	5.3	27.9	48.2
FS60-I	280(S)	186(I)	1088	700	0.28	130	785	2206	150	5.2	25.5	41.5
FS50-III	233(S)	233(III)	1088	700	0.28	130	1429	2175	110	6.8	41.0	49.9
FW40-I	186(W)	280(I)	1088	666	0.31	300	1000	2206	125	5.7	34.3	50.7
FW50-I	233(W)	233(I)	1088	666	0.31	300	1437	2311	110	5.0	34.4	48.1
FW60-I	280(W)	186(I)	1088	666	0.32	450	1250	2284	150	6.0	21.9	37.6
FF40-I	186(F)	280(I)	1198	590	0.28	150	1083	2167	130	7.0	29.6	42.1
FF50-I	233(F)	233(I)	1198	590	0.28	200	1511	2208	100	5.4	33.5	49.6
FF60-I	280(F)	186(I)	1198	590	0.28	150	1408	2167	150	6.0	24.6	43.4
C-III	***	380(III)	1088	700	0.36	100	2500	2285	170	4.3	44.1	58.7
C-I	***	380(I)	1088	700	0.32	70	1406	2285	100	5.0	46.8	53.9

[a]S = Sundance, W = Wabamun, F = Forestburg. [b]I = ASTM Type I Cement, III = ASTM Type III Cement. [c]W = water, C = cement, F = fly ash. [d]AEA = air entraining agent (ml per 100 kg of cementitious material). [e]Superplasticizer (ml per 100 kg of cementitious material). ***Control mixes with no fly ash.

tion is relatively more pronounced than with other mineral admixtures. The presence of organic matter, specifically carbon content depicted in terms of loss on ignition, is normally believed to be the prime cause of the observed erratic behaviour of air entrainment in fly ash concrete. It is only through further research that the plausible reasons for variability of AEA demand in fly ash concrete can be determined.

Burns et al. (1982) have made attempts to neutralize the absorption characteristics of activated carbon in fly ash by using chlorine gas, calcium hypochlorite and some other surface active agents to resolve the problem of increased AEA demand. However, these preliminary studies need to be substantiated for practical use. The deactivating agents used for carbon in fly ash should not interfere with air entrainment and concrete durability on their own. Economy is another important factor to be considered in developing suitable remedial additives for solving the problem of AEA demand in concrete.

With the additional costs associated with special precautions, additional testing and inspection personnel required for quality control of fly ash and concrete to eliminate or minimize air entrainment problem etc, the possible economic benefits of fly ash usage in concrete may be defeated.

High carbon content in fly ash has been appreciably reduced by recent developments in the equipment for ash collection as well as power production; and high quality low carbon fly ash is presently being produced by several power companies for commercial use.

4.6. TEMPERATURE RISE IN FRESH CONCRETE

Hydration of Portland cement is accompanied by liberation of heat that raises the temperature of concrete. Because of the slower pozzolanic reactions, partial replacement of cement by

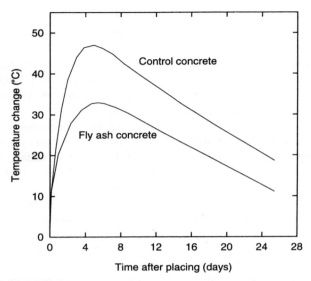

FIGURE 4.3   Temperature rise curve for fly ash and plain concrete test sections (Compton 1952).

fly ash results in release of heat over a longer period of time and the concrete temperature remains lower because heat is dissipated as it is evolved as indicated in Figure 4.3. In the most of the early uses of fly ash in concrete, the primary aim was to achieve reduction in the rise of temperature in fresh concrete. In mass concrete, where cooling following a significant temperature rise due to the generation of the heat of hydration occurs, stresses can develop and cause cracking. Temperature rise, in fact, depends upon more factors than the rate of heat generation associated with hydration and pozzolanic reactions, including the rate of heat loss and the thermal properties of the concrete and the surrounding medium. The size of concrete member plays a dominant role in temperature rise of a particular concrete as shown in Figure 4.4.

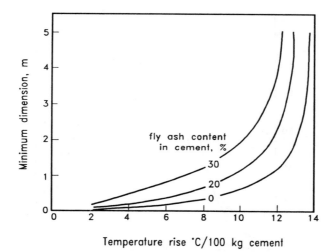

FIGURE 4.4   Effect of unit minimum size on the temperature rise in fly ash and plain concrete (William and Owens 1982).

Several investigators have reported favourable effects of fly ash incorporation in concrete on temperature rise of not only massive concrete dams, but also in concrete mat foundations and massive columns in lower storeys of tall buildings (Rodway et al. 1985, Sivasundaram et al. 1989b, Langley et al. 1989). Low calcium Class F fly ashes generally tend to reduce the rate of temperature rise more as compared to high calcium Class C fly ashes (Crow and Dunstand 1981). Some high calcium Class C fly ashes with self cementitious properties may react very rapidly with water, thus releasing excessive heat just like normal Portland cement hydration.

Bamforth (1980) in a study on fly ash concrete and slag concrete for use in large size foundation observed the temperature rise in concrete after placement as shown in Figure 4.5. It

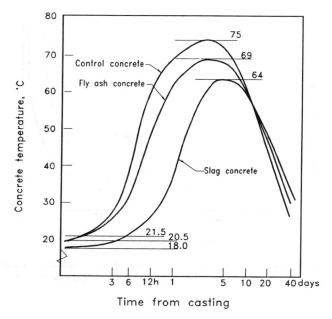

FIGURE 4.5   Variation of temperature recorded at mid-height in fly
ash, slag and plain concrete foundation units (Bamforth
1980).

is observed that with an increase in the quantity of cement
replaced by fly ash and slag, the rate of heat release is slowed
down and as a result the maximum temperature reached at any
point in the concrete mass is lower than the concrete containing
no fly ash.

ACI Committee 211.1-81 (1984) has estimated that on
equivalent mass basis fly ash contributes to early age heat liberation in the range of 15 to 30% compared to normal Portland
cement. Low calcium bituminous fly ash was successfully used
to control the rise of temperature at early age in the construction

of concrete structures, particularly dams (Philles 1967, Elfert 1973). In general, incorporation of fly ash as replacement of normal Portland cement exhibits less temperature rise than concrete without fly ash. Where early age strength is not the main design consideration, large quantities of cement replaced by fly ash are anticipated to reduce the rate and also amount of heat of hydration significantly. This of course is of great benefit in the mass concrete structures defined by ACI Committee 116 (1982) as "any volume of concrete with dimensions large enough to require that measures be taken to cope with generation of heat of hydration from the cement and attendant volume change to minimize cracking."

CHAPTER 5

Effects of Fly Ash on the Structural Properties of Hardened Concrete

5.1. GENERAL

This chapter deals with the effects of fly ash addition on the properties of hardened concrete such as strength, elasticity, shrinkage and creep, thermal expansion, permeability and durability of concrete against exposure to different environments including weather, chemical, nuclear and biological. The usual primary requirement of a good concrete in its hardened state is a satisfactory compressive strength, as the majority of the desired properties of hardened concrete are concomitant with high strength. Therefore more emphasis has been laid on the ways in which the use of fly ash influences strength development of concrete under different curing regimes, especially curing temperatures. Several comprehensive reviews as well as other publications summarize the accepted views concerning the strength development characteristics and other structural properties of concretes containing fly ash. A brief description of the effects of fly ash based upon several research studies and field applications concerning the performance of hardened concrete are presented in this chapter.

5.2. STRENGTH DEVELOPMENT

Fly ash concrete in which a portion of the cement normally used is replaced by a fly ash having adequate pozzolanic properties as defined in ASTM C618 or similar standards, will ultimately develop greater strength than the similar concrete without fly ash. However, the rate of strength development and the level of such strength depends on several factors including the characteristics of fly ash, the type of cement used, replacement level of cement with fly ash, mix proportions, ambient temperature and curing environment, and presence of other additives (Hobbs 1983a). Properties of the fly ash including its chemical and mineralogical composition, fineness, pozzolanic reactivity, and the temperature and other curing conditions are equally important variables which affect the strength and durability of concrete.

The general belief that fly ash concrete has low early strengths has emerged from the early research findings, involving mostly low calcium Class F fly ash that show comparisons of strength gains of two concretes containing the same aggregate. In one case Portland cement was the only binding agent and in the other a portion of the cement was replaced by Class F fly ash on a volume for volume or weight for weight basis. In addition, the ashes used in most of the early work were produced in older power plants and were not as fine, contained high carbon content and often exhibited much lower pozzolanic activity. With such ashes used in concrete as simple replacement of cement showed significantly lower strengths at early ages. Nevertheless, the concretes made with all types of ashes which possess pozzolanic characteristics usually develop strengths higher than those of similar concrete without fly ash at later ages.

When high calcium Class C fly ash is used, the strength development with time is likely to be different from that obtained using Class F fly ash. The self hardening reactions in the Class C fly ashes are likely to occur within the same time frame as the normal Portland cement hydration reactions, giving equal or sometimes greater strengths at early ages. The pozzolanic activity of such cementitious fly ashes further enhances strength at later ages. Several recent studies have demonstrated that very high strength concretes can be produced with some Class C fly ashes, having high calcium oxide contents, generated in modern power plants.

Both the rate of strength gain and the ultimate strength of concrete are of concern to the construction engineer. For most of the structural applications of concrete, the slow rate of strength development is of concern particularly when concrete is placed in cold weather or when high early strength is a structural requirement for load application, and transfer of stresses by early removal of formwork as in slip forming etc. To counteract potentially low early strength, an amount of fly ash in excess of the amount of cement replaced is added as explained in modified replacement method of mix proportioning of fly ash concrete in Section 3.2.3.

Adjustment in the amount of fine aggregates in concrete can also be made to achieve higher early and ultimate strength in fly ash concrete. When so proportioned, fly ash concrete should have adequate strengths at early ages to meet the structural requirements.

In cold climates, many countries do not permit fly ash concrete placement after a specific date in the fall and not before a specific date in the spring. However, recent studies have indicated that with proper mix proportioning and with appropriate cold weather construction practices the fly ash concrete can be placed under the same ambient condition as normal cement concrete and thus earlier "cut off" dates for fly ash concrete may not be necessary. Early age strength is generally not a design requirement in mass concrete construction where much of the early use of fly ash in concrete is to control the temperature rise due to reduced heat of hydration. Nonetheless, the pozzolanic reactions that occur at a slower rate, provide for equal or greater ultimate strength for such concrete with fly ash as compared to regular concrete without fly ash.

Class F bituminous fly ash contributes to the long term strength gain of concrete more than Class C sub-bituminous or lignite fly ash in spite of its relatively slower rate of strength development at early age. Swamy and Mahmud (1986) reported that concrete containing 50% low calcium bituminous fly ash, as cement replacement, and using a superplasticizer is capable of developing 60 MPa compressive strength at 28 days and strength of 20 to 30 MPa at 3 days. For a typical low calcium fly ash, it was found that the pozzolanic reaction started at 11 days after hydration at 20°C and the significant effect on compressive strength appeared to occur after 28 days of curing. Many investigators have reported that no significant contribution to strength development is noticed up to 7 days with the use of low calcium fly ash in concrete (Mehta 1994a,b). At 28 days and beyond most fly ashes at the replacement levels of up to 30% by cement weight exhibit strength gain in concrete and the strength generally equals that of control concrete.

For more than five decades, high strength fly ash concrete has found applications in major structures, for example, high rise buildings in the Chicago area of the United States and the Toronto area of Canada. High strength concrete containing 504 kg/m^3 Portland cement, 59 kg/m^3 low calcium fly ash, and water reducing agent with water to cementitious material ratio of 0.33, used in the Water Tower Place and River Plaza in Chicago, developed about 70 MPa compressive strength at 50 days (Berry and Malhotra, 1982). For field applications in Texas, similar strengths were obtained from concrete mixes made with 400 kg/m^3 cement, 100 kg/m^3 high calcium fly ash containing about 30% CaO, water reducing agent, and with water to cementitious material ratio of 0.33 (Cook, 1982).

Hague et al. (1988b) reported that Class F bituminous fly ashes from Alberta power plants can be used under proper curing conditions to produce concrete at 35 to 50% cement replacement having 80 to 100% of the strength of an equivalent plain concrete after being cured for 90 days. Langley (1988) conducted tests to determine the effect of a bituminous Class F fly ash from Eastern Canada on concrete with 50% replacement level of cement by fly ash. The results of this study presented in Table 5.1 indicate that flexural strength and splitting tensile strengths of fly ash concretes are slightly lower than those of the control mixes.

Hague et al. (1988b) reported that for concrete mixes with 40 to 75% bituminous fly ash replacing cement, the increase in flexural strength was slightly less than the increase in compressive strength between 28 days and 91 days of curing.

Compressive strength data of concrete containing 50% Alberta bituminous ash by weight as cementitious material are presented in Figure 5.1 from a paper by Joshi et al. (1983). Their results did not show any trend as no adjustments were made in water cement ratio and mix proportions to maintain constant workability of the mix. For all the mixes, the water cementitious ratio was kept at 0.47. All the fly ash mixes, however exhibited increase in strength beyond the age of 28 days. The data on these mixes in fresh state are presented in Table 4.2.

In a laboratory study, Joshi et al. (1993) tested a large number of fly concrete mixes made by using three different Alberta fly ashes containing about 10% calcium oxide. The replacement level varied between 40 to 60% by weight of cement. The mixes were superplasticized and air entrained to obtain 100 to 120 mm slump and $6 \pm 1\%$ air content. The cementitious material content varied from 380 to 466 kg/m^3, water to cementitious material ratio varied from 0.27 to 0.37, coarse aggregate ranged from

TABLE 5.1 Mechanical properties of hardened concrete (Langley 1989)

Cement type	Mix no.	$\frac{W}{C+F}$	Flexural strength (MPa)				Splitting tensile strength (MPa)		Modulus of elasticity (GPa)	
			14 d	28 d	91 d	365 d	28 d	365 d	28 d	365 d
I	0	0.20	–	–	–	–	–	–	–	–
	1	0.28	7.6	–	9.6	–	–	–	36.1	–
	1C⁺	0.39	8.0	–	–	–	–	–	31.5	–
	2	0.30	–	6.9	–	9.0	4.2	9.0	35.1	46.6
	2C⁺	0.39	–	7.8	–	8.3	4.1	4.0	36.8	38.2
	3	0.33	6.5	–	8.9	–	–	–	33.8	–
	3C⁺	0.45	7.0	–	7.4	–	–	–	28.9	–
	4	0.35	–	6.0	–	7.5	3.8	4.8	31.6	34.5
	4C⁺	0.46	–	7.4	–	7.0	4.1	3.2	34.9	34.4
	5	0.49	–	–	*	*	2.9	*	27.9	*
III	6	0.30	–	7.3	–	9.2	4.1	5.5	32.7	46.1
	6C⁺	0.39	–	8.6	–	7.0	4.7	4.2	32.2	36.3
	7	0.35	–	6.4	–	8.0	3.3	4.5	32.1	43.1
	7C⁺	0.46	–	7.0	–	6.4	3.8	3.5	32.6	34.8

– = test was not performed. *Results not yet available. ⁺Control mixes with no fly ash. All other mixes contained 50% fly ash with respect to total cementitious material.

1012 to 1194 kg/m³, and fine aggregate or sand varied from 712 to 643 kg/m³. A summary of the mix proportions and other relevant test data along with 7 days and 28 days compressive strengths of a few typical mixes are presented in Table 4.4. The results indicated that with fly ash replacement levels up to 50% by weight of cement, concrete with 28 day strength ranging from 40 to 60 MPa could be produced.

At 7 days, the fly ash concretes had strength ranging between 27.9 MPa and 41.0 MPa compared to 44.1 MPa of control concrete. However at 28 days, the fly ash concretes developed strength varying from 37.6 to 50.7 MPa against 58.7 MPa for reference plain concrete. Nevertheless after 120 days, the observed strength of fly ash concrete in the range 54.8 MPa to

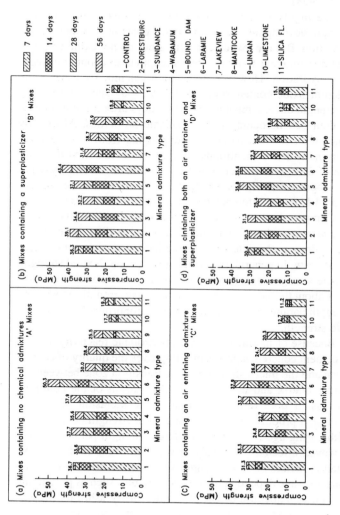

FIGURE 5.1 Summary of compressive strength results of various fly ash concrete mixes (Joshi et al. 1987).

74.6 MPa was quite close to 74.6 MPa of control concrete (Joshi et al. 1993).

The highly pozzolanic fly ashes start their contribution to strength development almost from the onset of Portland cement hydration. But the low calcium fly ashes do not exhibit significant pozzolanic activity to affect strength until about two weeks after hydration. However, some high calcium fly ashes, with calcium oxide content more than 15%, may start contributing to strength development as early as 3 days after mixing because of their self hardening and pozzolanic properties. Alberta fly ashes with calcium oxide content ranging from 7 to 13% are marginally self cementitious but quite pozzolanic. At the early age of 3 days the concrete mixes containing Alberta fly ashes developed relatively less strength of 13.7 to 28.6 MPa compared to 35.3 MPa for reference concrete (Joshi et al. 1993).

Compressive strength usually is good indicator of the quality of concrete with and without fly ash and is directly related to the hydration characteristics of the hardened cementitious material paste. In general, the effect of fly ash on flexural strength, tensile strength, and bond strength of concrete with steel, follows about the same pattern as compressive strength. Most other strength parameters are approximately proportional to compressive strength.

Because of its fineness as well pozzolanic reactivity, fly ash in cement concrete significantly improves the quality of cement paste and the micro structure of the transition zone between the binder matrix and the aggregate. As a result of the continual process of pore refinement, due to the inclusion of fly ash hydration products in concrete, a gain in strength development with curing age is achieved. Different strength parameters are, however, affected favourably in varying magnitudes. The correlation factors of other strength with compressive strength at a particular age depend upon several variables such as pozzolanic as well as cementitious characteristics of fly ash, mix proportions and curing conditions including initial temperature.

5.2.1. Effect of Curing Conditions

In order to promote continued hydration and pozzolanic reactions in the concrete containing fly ash, favourable conditions of temperature and humidity are necessary. Poor curing conditions affect the strength of fly ash concrete, in particular the one with a high proportion of ash, much more than the strength of plain concrete. Several investigations have been undertaken to study the effects of moist curing or fog curing, water curing, dry curing or air drying, and mass curing at normal temperature on properties of hardened concrete such as strength, permeability, volume stability, and durability.

Hague et al. (1988b) reported that mixes with Alberta fly ashes replacing up to 50% cement show smaller reduction in strength at lower ash contents when curing is done at 50% relative humidity at room temperature of about 23°C. The effect of this poor curing was not apparent, however, before 8 days because the test specimens contained sufficient residual water for hydration to continue. Gifford et al. (1987) studied the effect of dry curing on unprotected concrete specimens at 0 and 5°C made with 40% cement replaced by the same type of fly ash as studied by Hague et al. They found that within 50 hours after casting, about 30 and 60% of the mixing water was lost through evaporation from unprotected surfaces on curing at 50 and 10% relative humidity, respectively. In addition to strength loss due to evaporation of water, it was reported that the combination of low curing temperature and the cooling effect of concomitant evaporation of water at low humidity would significantly retard the rate of hydration and thereby strength development. The importance of curing concrete in an enclosed environment particularly at low temperatures, is therefore stressed in order to mitigate the effect of water evaporation during the initial hours after the placing of concrete.

Based on the satisfactory level of strength development at 28 days for a high volume fly ash concrete, made with Class F fly ash that had been moist cured for 3 days, Langley et al. (1989)

TABLE 5.2   Rate of strength development as percentage of 28-day strength (Swamy and Mahmud 1986)

Age (days)	Strength (MPa)								
	20			40			60		
	FOG	DRY	7F+D	FOG	DRY	7F+D	FOG	DRY	7F+D
1	19	–	–	25	–	–	34	–	–
7	55	71	–	67	80	–	68	79	–
28	100	100	100	100	100	100	100	100	100
150	176	125	104	140	119	105	135	123	117
270	197	114	110	150	125	116	140	113	121
365	209	118	106	162	122	107	146	116	118

7F+D = 7 days fog followed by air dry curing.

reported that minimum duration of moist curing for fly ash concrete was 3 days after which normal curing practices as for ordinary plain concrete might be employed without any significant adverse effects. They also pointed out that long term strength development in mass fly ash concrete was less influenced by dry curing than much smaller test specimens used in the laboratory.

The effect of curing regime on strength development of high volume fly ash concrete mixes can be estimated from the data in Table 5.2 (Swamy and Mahmud 1986). Increase in strength of 50 to 100% over the 28 day strength of fly ash concrete was achieved after one year under continuous moist or fog curing compared to only 18 to 25% increase for the control or plain concrete under similar curing conditions. Under the other two curing regimes, one with 7 day moist curing followed by air drying and the other with continuous dry curing, the corresponding increase in strength of fly ash concrete after one year varied between 6 to 22% of the 28 day strength of the reference concrete.

5.2.2. Effect of Curing Temperature at Early Age on Strength Development

The rate of pozzolanic reaction of fly ash in cement concrete is significantly influenced by curing temperatures at early ages as is the case for cement hydration reactions. Pozzolanic reactions are highly temperature dependent. The higher the curing temperature the higher rate of the pozzolanic reactions. However it has been reported that at curing temperatures in excess of 30°C, fly ash concrete behaves in a significantly different manner than concrete made with Portland cement only as shown in Figure 5.2. In case of control concrete cured at elevated temperatures, the rate of strength gain increased at early ages but a noticeable decrease in strength gain occurred at later ages. On the contrary, in case of fly ash concrete gain in strength is achieved monodically as a consequence of heating. It is believed that the products of fly ash and cement hydration, their relative proportions and microstructure, are markedly different from those formed from thermally activated hydration of Portland cement alone.

The favourable effects of fly ash in concrete cured at moderately elevated temperatures can be advantageously used in the construction of mass concrete or in precast concrete operations involving accelerated curing at elevated temperatures. In the construction of an intake tunnel of the Kurabegwa No. 3 Power Station in Japan (1995, 1996), concrete with 25% cement replaced by fly ash was used to offset the loss of strength due to elevated rock temperatures of 100 to 160°C that would have occurred had plain Portland cement concrete been used.

Ravina (1981) reported that fly ash concrete subjected to high temperature at an early age of curing exhibited increased rate of strength gain possibly due to the higher rate of pozzolanic reactions. It was suggested that when concrete is cured at elevated temperatures large quantities of fly ash may be incorporated with a significant improvement in strength compared to the rather limited contribution under normal curing conditions up to 28 days. It was further suggested that the pozzolanic

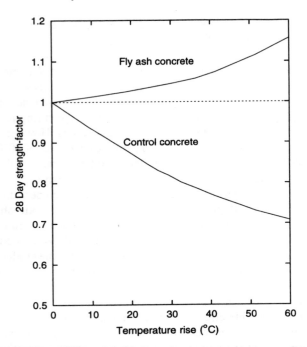

FIGURE 5.2   Effect of temperature rise during curing on compressive
strength development of concretes (William and Owens
1982).

reactions once triggered thermally appear to continue when the
source of external heating is removed. The beneficial effects of
curing at elevated temperatures manifest in considerable gain in
early age strength which increases progressively at later ages.

At low temperatures, reduction in the rate of strength development is caused by the combined effect of slower hydration of
Portland cement and retarded pozzolanic reactions. The possi-
bility that in some cases fly ash concrete may be affected by
cold weather in a manner different from regular concrete must
be considered. Generally fly ash concretes require more atten-

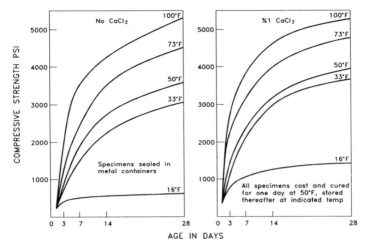

FIGURE 5.3. Compressive strength of concrete cured 24 hours at 50°F and stored at various temperatures (Price 1951).

tion in proportioning as well as curing when they are to be placed at cold temperatures. The effect of cold temperatures at very early ages on strength development is illustrated in Figures 5.3–5.6.

5.3. ELASTIC PROPERTIES

Several investigators (Lane and Best 1982, Lohtia et al. 1976, Ghosh and Timusk 1981) have observed that the effect of fly ash addition as cement replacement on modulus of elasticity of concrete is almost the same as on compressive strength. The modulus of elasticity of fly ash concrete is generally lower at an early age and is slightly higher at late ages. In their study Lohtia et al. (1977) found that as compared to compressive strength gain, the increase in modulus of elasticity was less with the incorporation of 15 to 25% Class F fly ash in concrete at

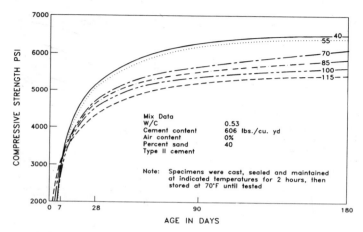

FIGURE 5.4. Effect of initial temperature on the compressive strength of concrete (Price 1951).

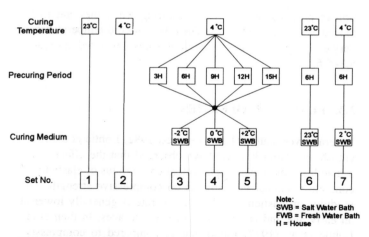

FIGURE 5.5. Cement paste specimen description (Ramachandran et al. 1990).

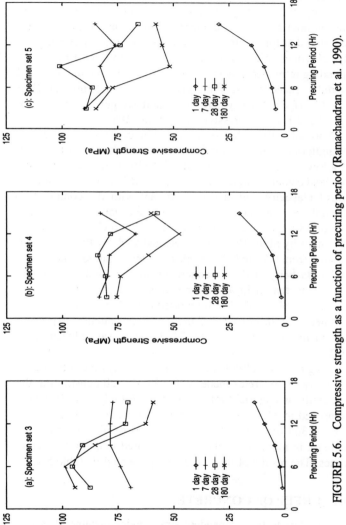

FIGURE 5.6.   Compressive strength as a function of precuring period (Ramachandran et al. 1990).

the age of 90 days. In general, fly ash increases the modulus of elasticity of concrete when concretes of the same strength with and without fly ash are compared. Ghosh and Timusk (1981) reported that for all strength levels the modulus of elasticity of fly ash concrete was generally equivalent to that of the corresponding reference concrete. They also found that the observed modulus exceeded the value given by the ACI formula, $E_c = 0.043 \, W \, (f_c)^{3/4}$ MPa, where W is unit weight of concrete in kg/m^3 and f_c is compressive strength in MPa.

Normally fly ash properties affecting the compressive strength of concrete also influence the modulus of elasticity but to a lower extent. Crow and Dunstan (1981) reported that like Portland cement concrete, fly ash concrete had increased modulus of elasticity with age concomitant with the compressive strength development. The modulus of elasticity ranged from a low of 18.8 GPa at 28 days to a high of 39.6 GPa at 365 days. The majority of the fly ash concrete had a 28 day Poisson's ratio ranging from 0.14 to 0.25. At elevated temperatures, the modulus of elasticity of fly ash concrete using Saskatchewan lignite fly ash decreased in a similar way as that of plain concrete. Nasser and Marzouk (1983) reported that when fly ash concrete was heated from 21 to 232°C in the sealed containers to prevent loss of moisture, the modulus of elasticity was reduced up to 40%.

Langley et al. (1989) found that at 28 days the modulus of elasticity of concretes made with 50% fly ash constituting the cementitious material varied between 27.9 GPa and 36.1 GPa compared to 31.5–36.8 GPa for control concrete mixes. However at 365 days fly ash concrete mixes exhibited significant increase in modulus of elasticity compared to control concrete mixes. The results of their study are presented in Table 5.1.

5.4. CREEP OF CONCRETE

Creep refers to time dependent strain under sustained loading. The effects of fly ash on creep of concrete are limited primarily to the extent to which fly ash influences the ultimate strength and rate of strength gain. Because of the increase in strength with age due to pozzolanic reactivity, concrete with fly ash proportioned to have the same strength at the age of testing as concrete without fly ash would produce less creep strain at all subsequent ages. Relatively few studies have been conducted on creep of fly ash concrete. Lohtia et al. (1976) reported that cement replacement of 15% by Class F Indian fly ash had insignificant effect on creep of concrete. With increase in replacement level from 15 to 25%, fly ash concrete exhibited slightly higher creep than the equivalent plain concrete. Creep recovery for fly ash concrete with 15% fly ash ranged from 23 to 43% of the corresponding 150 day creep. For higher fly ash replacement levels, creep recovery was found to be still smaller. The nature of creep time curves for plain and fly ash concrete mixes as shown in Figure 5.7 are similar. In the reported study, cement replacement of 15% by fly ash was found to be optimum since it improved strength, elasticity, shrinkage and creep characteristics of structural concrete as shown in Figure 5.8.

In an investigation on concretes made with bituminous fly ashes of different carbon contents and fineness values, Ghosh and Timusk (1981) found that fly ash concrete proportioned for equivalent 28 day strength of plain concretes ranging from 20 to 55 MPa showed less creep than the plain concrete. The probable reason for the observed behaviour was the favourable effect of pozzolanic reactions due to fly ash which caused higher rate of strength gain after the time of loading than for the plain concrete. The results of study by Yuan and Cook (1983) on high strength concretes containing 20 to 50% high calcium fly ash showed opposite trend in that fly ash concrete containing 30 and 50% ash had more creep than control concrete. However, the 20% fly ash concrete exhibited about the same creep as control concrete as indicated in Figure 5.9. Gifford and Ward (1982) also reported that addition of fly ash in lean mass concrete reduced the creep. This was attributed to the increase in modulus of elasticity as well as reduction in the

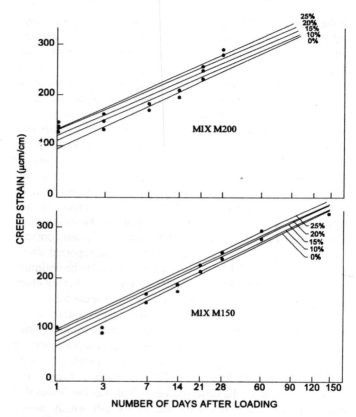

FIGURE 5.7. Creep strain versus time (log scale) under stress–
strength ratio of 0.2 for indicated mixes and fly ash
content, with the data corresponding to 15% cement
replacement superimposed (Lohtia et al. 1976).

relative volume of paste available for creep, when fly ash was
incorporated in lean mass concrete.

Nasser and Al-Manasser (1986) conducted creep tests on
superplasticized and air entrained concretes containing 20 to
50% lignite fly ash from Saskatchewan. They found that con-

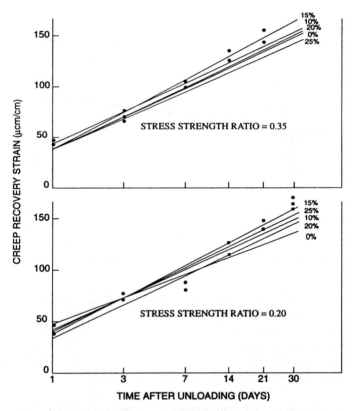

FIGURE 5.8. Creep recovery strain versus time (log scale) for mix M150 at indicated fly ash content with observed data of 15% fly ash superimposed (Lohtia et al. 1976).

crete samples containing 20% of fly ash in unsealed conditions had on average 72% greater creep than the equivalent sealed concrete for stress–strength rates varying from 10 to 60%. However concrete with 50% fly ash had 13% less creep for unsealed specimens and 39% less for sealed ones. Similar behaviour was observed in the tests conducted by Carette and Malhotra (1987)

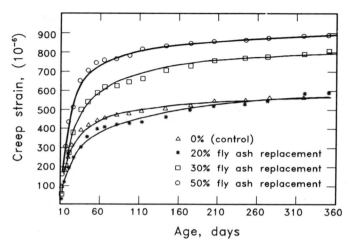

FIGURE 5.9.   Creep of fly ash concretes (Yuan and Cook 1983).

with various types of Canadian fly ashes including Saskatchewan fly ash. In the study concrete with 20% fly ash produced
consistently lower creep compared to that of control concrete.
Nasser and Marzouk (1983) had also observed that creep of
concrete with 20% fly ash under sealed conditions showed a
continuing reduction with an increase in temperature except at
177°C, while that of unsealed specimens increased with temperature up to 71°C and decreased thereafter within the range
of temperature between 21.4 and 232°C.

Since creep is influenced by compressive strength and modulus of elasticity of concrete, higher creep strains are observed
in fly ash concrete at early age loading when strength is low.
However, the creep rate decreases at later ages. Creep of concrete is generally related to the removal of adsorbed water from
hydrated paste and to the viscosity of the paste. With the progressive hydration and pozzolanic reactions in fly ash concrete,
compressive strength increases due to the increase in the amount

of cementitious hydration products and refinement of pore structure. The increase in compressive strength of concrete beyond the age of loading reduces creep under sustained stress as stress-to-strength ratio is relatively reduced.

5.5. LOAD INDEPENDENT VOLUME CHANGES

Volume changes in concrete due to hydration reactions, drying, wetting and drying, and thermal variation occur even without application of any loads. Under normal field conditions, drying shrinkage is an important phenomenon which induces time dependent deformations and is sometimes called creep under zero load.

The majority of the investigations (Elfert 1973, Munday et al. 1982, Nasser and Al-Manasser 1986) have reported that for replacements of up to 20% of cement with fly ash, no significant difference occurs in drying shrinkage in properly proportioned mixes. Increase in drying shrinkage with fly ash addition may occur from increases in the paste volume if water content is the same. However if water content is reduced, shrinkage is minimal. Munday et al. (1982) reported that incorporation of fly ash does not significantly affect the shrinkage and expansions due to drying, wetting/drying and thermal changes in concrete. Ghosh and Timusk (1981) reported that for the same maximum size of aggregate and for all strength levels, the shrinkage of concrete containing fly ash was lower than that of concrete without fly ash.

Yuan and Cook (1983), from their studies of concretes made with high calcium fly ash, concluded that the replacement of cement in the range of 20 to 50% by fly ash has insignificant influence on drying shrinkage as shown in Figure 5.10. However, the test data by Hague et al. (1988b) on concrete with 40 to 75% cement replacement by a bituminous fly ash containing 10% calcium oxide indicated that drying shrinkage of concrete decreased with increase in fly ash content.

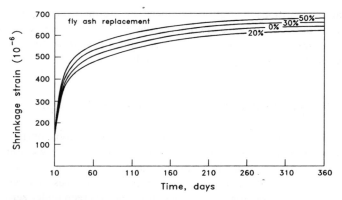

FIGURE 5.10. Drying shrinkage of concretes incorporating high-calcium fly ash (Yuan and Cook 1983).

In a study by EPRI (1987) involving a wide range of American fly ashes in about 177 concrete mixes the effect of fly ash on concrete properties was investigated. The studies were conducted at five different testing laboratories to relate the performance of concrete to the physical/chemical characteristics of fly ashes collected from 16 power plants across the United States. Fifteen to 60% of the cement was replaced by fly ash in concrete. Tests were conducted to determine compressive strength, modulus of elasticity, flexural strength, shrinkage, and freeze–thaw durability. Shrinkage was determined in terms of percent change in length from 1 day old specimen in accordance with ASTM C157. The specimens were cured in water for 28 days and then subjected to air curing for additional 62 days. No significant difference in shrinkage was observed in fly ash concrete mixes as compared with the corresponding control concrete mixes. From the majority of the published literature, it is evident that incorporating fly ash in concrete mixes, in general, has insignificant influence on the load independent volume change including drying shrinkage of the concrete.

CHAPTER 6

Effects of Fly Ash on the Durability of Concrete

6.1. PERMEABILITY

Concrete must be relatively impervious so as to enable it to withstand the service conditions for which it has been designed, without serious deterioration over the life span of the structure. The loss of concrete durability may be caused by the severity of the environment to which it is exposed or by internal changes within the mature concrete itself. The external causes may be physical, chemical or mechanical such as weathering, extremes of temperatures, abrasion and erosion, chemical action, and attack by natural or industrial aggressive liquids and gases. Some of the internal causes include alkali–silica reaction and incompatible dimensional changes due to the differences in the thermal properties of aggregate and cement paste. The present state of knowledge has established, that of all the causes which affect concrete durability, the main one is the permeability to gases, liquids and salts. Less permeability generally means more durability.

Based on several laboratory investigations and field observations, it is found that at early ages fly ash concretes, containing low or high calcium fly ashes at replacement levels of up to 50% as cementitious material, are more permeable than plain concretes containing no fly ash. This trend is found to reverse

111

after about 180 days most likely because of pozzolanic activity of fly ash in concrete (Davis et al. 1954).

The permeability of concrete depends primarily on the size, distribution and continuity of the pores of the hydrated paste of the concrete. The two important factors which control the pore structure of the paste are degree of hydration and water cement ratio. The addition of fly ash can cause considerable pore refinement i.e. transformation of bigger pores into smaller ones due to the formation of pozzolanic reaction products concomitant with the progress of cement hydration. Since strength and impermeability are inversely related to the volume of pores larger than 100 Å in the hydrated paste, the phenomenon of pore refinement in fly ash concrete leads to the improvement in these characteristics.

In ordinary Portland cement concrete, calcium hydroxide formed during hydration of Portland cement can be leached out over a period of time. This creates channels available for the ingress of water and deleterious salt solutions. However, when fly ash is added, it reacts with the calcium hydroxide in the water filled capillary channels to produce calcium silicate and aluminate hydrates of the same or similar type that are formed in the normal hydration of cement. Thus the calcium hydroxide is consumed in the pozzolanic reactions and converted to water insoluble hydration products. The reactions reduce the risk of leaching calcium hydroxide. The reaction products also tend to fill capillaries, thereby reducing permeability to aggressive fluids such as chloride or sulphate solutions.

In the reported studies (Kanitakis 1981) on permeability of concrete with fly ashes containing less than 2% calcium oxide, it was concluded that at early ages fly ash concrete behaves as a lean mix concrete and is thus permeable. However, with the progress of pozzolanic reactions the permeability is reduced at later ages.

Diffusion of ions, such as chlorides in cured concrete is generally represented by Fick's diffusion law (Barrer 1951):

$$\frac{dc}{dt} = D_c \frac{d^2c}{dx^2},$$

where c is the chloride ion concentration at a distance x after time t, and D_c is the diffusion coefficient of chloride ions in concrete.

The value of D_c may vary from 1.0×10^{-9} to 50×10^{-9} cm^2/s for high and normal strength concrete, respectively. Short and Page (1982) reported that D_c value for paste with fly ash Portland cement was 14.7×10^{-9} cm^2/s compared to 44.7×10^{-9} cm^2/s for normal Portland cement paste. This led to the conclusion that fly ash concrete was more effective in limiting chloride diffusion than Portland cement concrete.

Kasai et al. (1983) studied the air/gas permeability of mortar made with blended cements containing fly ash and blast furnace slag. The study was important from durability aspects of concrete with respect to carbonation. At early curing ages of up to 7 days, the blended cement mortars exhibited more permeability than plain cement mortars. However, with increased curing age, the permeability of blended cement mortars decreased. In general terms, the permeability was found to be directly related to the compressive strength development of the mortars.

In general an improvement in impermeability and thereby durability of concrete against frost attack, sulphate attack, alkali aggregate reaction, and also resistance of concrete against the reinforcing bar corrosion, is significantly improved with the incorporation of fly ash in concrete. These favourable effects of fly ash on various aspects of concrete durability are described next.

6.2. FREEZE–THAW DURABILITY

As is the case with all concretes in general, the resistance of fly ash concrete to damage from freezing and thawing depends on the adequacy of air void system, the soundness of aggregate, degree of hydration and strength of binding paste, maturity and moisture content of concrete. Freeze–thaw durability is evaluated using the test procedure ASTM C666. Of course as explained earlier in Section 4.5, special attention must be given to attaining the proper amount of entrained air when fly ash is used in concrete.

Some researchers have reported that even with adequate entrained air, fly ash concrete has a lower frost resistance when compared to that of concrete without fly ash at equal ages. However, when comparison is made under conditions that the fly ash concrete has developed adequate strength equal to that of reference concrete, no significant differences in freeze–thaw durability have been observed provided the air void system remains constant.

Air entrainment has the greatest influence on freeze–thaw durability of both fly ash and plain concrete mixes. Gebler and Klieger (1983, 1986b) conducted tests to investigate the influence of air entrainment on the air–void parameters of hardened concretes made with both Class C and Class F fly ashes. The concretes were cast after 30, 60, and 90 minutes of initial mixing. They found that air void spacing factors were almost constant for the majority of concretes containing fly ash which were cast after 90 minutes. At early periods of casting concretes with Class F fly ash exhibited greater variability in air void parameters than concretes with Class C ash.

The test results on freeze–thaw durability of concretes made with low calcium fly ashes containing 5.4, 12.3 and 23% carbon, and at 15, 30, 45 and 60% replacement level by cement weight showed that the correlation between durability factor as determined by resonant frequency and carbon content of fly ash was poor (Sturrup et al. 1983). The concrete mixes had water to cementitious material ratio of 0.6 and air content of 6.5 ± 1%, except for the 23% carbon fly ash concrete specimens which had air content of 3.5%. The specimens when exposed to outdoor freezing–thawing attack showed more definite weight loss which indicated surface scaling.

Yuan and Cook (1983) studied freeze–thaw performance of both air entrained and non-air-entrained concretes made with high calcium fly ash at 0, 20, 30 and 50% levels of substitution for cement by weight. They found that freeze–thaw resistance significantly improved by air entrainment in both plain and fly ash concretes. Furthermore, concrete with 20% fly ash exhibited better frost resistance than the control concrete. However, when fly ash level in air entrained concrete was increased to 50%,more scaling damage was observed after 400 cycles of freezing and thawing.

Carette and Malhotra (1984) conducted freeze–thaw durability tests on concrete mixes with a wide range of Canadian fly ashes. With constant air content around 6.5%, all the fly ash concretes have about the same durability factor as control concretes. Larson (1964) and Virtamen (1983) reported a similar behaviour that addition of fly ash showed no significant effect on frost resistance of concrete when strength and air content were kept constant. Freeze–thaw durability tests when conducted after long curing periods have indicated that with the incorporation of fly ash, as is also the case with the use of blended cements in concrete, the resistance to frost attack significantly improves due to the development of strength equivalent to or superior to those of ordinary Portland cement concretes.

Freeze–thaw tests conducted by Joshi et al. (1987), on concrete with 50% cement replaced by Alberta Fly ash, exhibited relative dynamic modulus of elasticity values in excess of 60% after 300 cycles. No significant differences were observed in freeze–thaw performance of air entrained high volume fly ash concrete and the concrete containing both air entraining and water reducing agents. However, the test specimens without air entrainment failed at less than 50 freeze–thaw cycle indicating a low level of durability. They also reported that fly ash concrete mixes showed some scaling after 150 to 200 freeze thaw cycles and exhibited about 2% weight loss at the end of the 300 cycles.

Gifford et al. (1987) reported that, with the use of the same Alberta fly ash at 40% replacement level by cement weight, the concrete employed in curb and gutter construction with air content of around 6.5% provided freeze–thaw resistance similar to that of control concrete with no fly ash.

In a recent study, Joshi et al. (1993) conducted freeze–thaw tests on air entrained and superplasticized concretes made with 40 to 60% Alberta fly ash as cement replacement to produce high strength concrete mixes at water to cementitious material ratios of 0.28 to 0.36. They reported that most of the mixes had relative dynamic modulus of elasticity values in excess of 60% after 300 freeze–thaw cycles. The specimens were moist cured for 14 days prior to freeze–thaw tests. The air content of the mixes was maintained at $6 \pm 1\%$. The use of fly ash in concrete increased the air entraining agent demand to maintain constant air content. In general, all the concrete mixes containing Alberta fly ash and with air content more than 5% exhibited acceptable freeze–thaw durability performance.

The majority of the concrete specimens subjected to freeze–thaw durability tests including the control specimen showed slight to moderate scaling after 200 to 250 cycles and exhibited weight loss of $0.5 \pm 0.1\%$ at the end of the test. The relative dynamic modulus of elasticity values of the most of the fly ash concrete mixes tested for freeze–thaw resistance were in excess of 75% after 300 cycles as can be seen in Figures 6.1 and 6.2. It may be noted that the specimens were only 14 days old at the start of the test and less than 90 days old when the tests were terminated after the stipulated 300 freeze–thaw cycles. With increase in curing age beyond 14 days and also with a relatively slower pace of freeze–thaw attack under actual field conditions, it was concluded that the design mixes with high fly ash content would possess similar durability against frost attack as ordinary Portland cement concrete provided adequate air was entrained.

Schiepl and Hardtle (1994) reported that the main effect of pozzolanic reactions due to fly ash addition is to change the pore

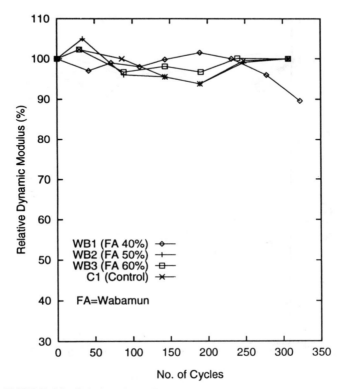

FIGURE 6.1. Relative dynamic moduli versus number of freeze–
thaw cycles for Wabamun fly ash concrete mixes (Joshi
et al. 1991).

size distributions, the total porosity remaining mostly un-
changed. In particular, the capillary water along with calcium
hydroxide liberated during cement hydration is largely consumed in pozzolanic reactions. The phenomenon of pore refinement as a result of pozzolanic reactions leads to the breaking of
continuity of the capillary pore structure. Several researchers

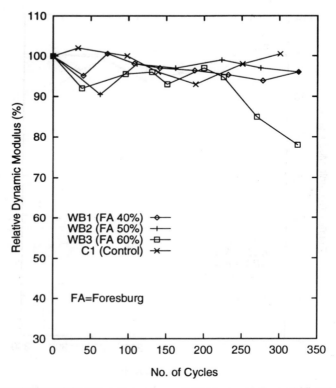

FIGURE 6.2. Relative dynamic moduli versus number of freeze–
          thaw cycles for Forestburg fly ash concrete mixes
          (Joshi et al. 1991).

have reported that the pozzolanic reactions alter the pore structure of the cement paste and thus density the transition zone
between the paste and aggregates (Malhotra et al. 1982).

The freezing temperature of water in small capillary pores
formed by fly ash addition is reduced and thus freeze–thaw
durability of concrete at some intermediate low temperatures is
improved. With the addition of fly ash, the water tightness of

the cement paste is also enhanced and thus the rate of water penetration is reduced. This reduction in water ingress rate, together with the decreased freezing temperature of capillary pore water and the reduction of the degree of saturation improves the resistance of fly ash concrete to frost attack. On the debit side, the reduced permeability of hardened cement paste with fly ash addition can retard internal moisture migration through the cement matrix whereby high internal pressure can develop that may cause cracking and deterioration of concrete.

To date, no satisfactory knowledge of the relationships linking the mechanism of frost attack and the micro structure of the cement paste exists. In the study by Schiepl and Hardtle (1994), no clear correlation was found between frost resistance and change in the pore structure of concrete caused by fly ash addition. They reported that frost resistance to scaling of fly ash concrete is dependent on the changes in the pore structure. Furthermore fly ash concrete has been found to exhibit greater surface scaling than plain cement concrete in many laboratory studies as well as experimental installations under freezing–thawing environmental. However, such scaling does not affect the internal structure and integrity of the concrete when used in actual field conditions which are not as severe as experimental exposure conditions. Thus the general belief based upon several studies suggests that the effect of fly ash addition on freeze–thaw durability of concrete is insignificant provided strength and air content are maintained constant.

6.3. RESISTANCE TO AGGRESSIVE CHEMICALS

The loss of durability of concrete by chemical attack can be either due to the decomposition of cement paste or due to the disruptive internal expansion caused by chemical reactions in the paste or by combination of both the actions. Deleterious chemicals such as acid solutions can react with $Ca(OH)_2$ to form water soluble salts that can be leached out of the concrete over a period of time, thereby increasing the permeability of concrete and aggravating the damage by increased and faster ingress of harmful chemicals. Sulphates can react with $Ca(OH)_2$ and calcium aluminate compounds in concrete to form gypsum and calcium sulpho-aluminate, ettringite, that can cause internal disruption of the concrete by concomitant volume increase of the paste.

6.3.1. Resistance to Sulphate Attack

One of the primary benefits of using fly ash in concrete is the increased resistance to attack from sulphates and potentially corrosive salts that penetrate into the concrete and cause steel corrosion with accompanying cracking and spalling of the concrete. It is well known that the reaction of fly ash with calcium hydroxide released during cement hydration results in the formation of additional calcium alumino-silicate hydrates and accompanying reduction in permeability of the concrete.

Larsen (1985) and several other investigators have reported that the change in the pore structure of cement paste as a result of fly ash addition can not be the sole reason for the observed favourable performance of fly ash concrete subject to chemical attack, particularly sulphate attack. They observed the favourable effect of the fly ash on sulphate resistance of concrete mixes even at ages when the pozzolanic reaction of fly ash was still not particularly high. Accordingly, it is suggested that additional chemical–mineral interaction including the reduction in tricalcium aluminate, C_3A, and calcium hydroxide content contribute to sulphate resistance. The fly ash may combine with some alumina phases such as C_3A in the cement during the first few days of cement hydration to form primary ettringite, thus reducing the potential for expansive sulphate–alumina reactions responsible for sulphate attack. With the increased conversion of C_3A in the initial stages, less of the aluminate containing phase is available for reaction during subsequent sulphate attack.

Since the early studies by Davis et al. (1937) several other studies have been conducted to investigate the sulphate resistance of cement mortar/concrete containing different types and combinations of fly ash and cement. Dikeou (1970) studied the resistance to sulphate attack of a number of concrete mixes using three different types of Portland cement, namely Type I, II and V, three Portland fly ash cements, and twelve different type F fly ashes. He reported that all of the fly ashes greatly improved sulphate resistance although the improvement seemed to vary with type and amount of cement and also the fly ash. Based on his study, the following order of resistance from the most resistant to the least resistant for the different cements and fly ashes is reported.

> Types V plus fly ash
> Type II plus fly ash
> Type V
> Type II
> Type I plus fly ash
> Type I

Dunstan (1976, 1980) reported the results of a five year study on sulphate attack of concrete mixes made with lignite and sub-bituminous fly ashes. He concluded that lignite and sub-bituminous Class C fly ash generally reduced sulphate resistance when used in normal proportion. Dunstan reports that as the calcium oxide in the fly ash increases above a lower limit of 5% and the ferric oxide (Fe_2O_3) decreases sulphate resistance is reduced. He proposed the use of an indicator R defined as follows:

$$R = \frac{\% \, CaO - 5}{\% \, Fe_2O_3}$$

For the fly ashes used by Dunstan (1980), those having R values of 1.5 or less generally improved sulphate resistance while those with higher values did not. The general applicability of the R

factor to predict sulphate resistance of all fly ashes is required to be verified by further studies. With the use of fly ash at 25% cement replacement level, the sulphate resistance of concrete made with ASTM Type II cement at 0.45 water cement ratio has been related to the R factor as follows:

R limits	Sulphate resistance
<0.75	Greatly improved
0.75 to 1.5	Moderately improved
1.5 to 3.0	No significant change
>3.0	Reduced

Mather (1982) reported the data from the studies at the laboratories of the U.S. Corps of Engineers on sulphate resistance of concretes using 3 different types of cement with C_3A contents of 9.4, 13.1 and 14.6% and 10 different types of pozzolans including volcanic glass, condensed silica fume, and different types of fly ashes. The pozzolans were added in mortars as 30% replacement of cement by volume. The mortar specimens at the time of their exposure to 0.352 M Na_2SO_4 solution had reached about equal maturity measured in term of compressive strength of companion mortar cubes. The order of resistance for various pozzolans from the most to the least effectiveness was found to be as follows: condensed silica fume, volcanic glass, sub-bituminous fly ash, bituminous fly ash, and lignite fly ash.

A summary of the test results given by Mather (1982) states: "What seems to be suggested (by the results) is that a pozzolan of high fineness, high silica content and highly amorphous silica is the most effective pozzolan for reducing expansions due to sulphate attack on mortars with non sulphate resisting cements.... The pozzolans that resulted in poor performance were in 6 of 7 cases fly ashes produced by the combustion of lignite."

Joshi et al. (1987) reported that with the incorporation of Alberta fly ash at 15% replacement level by weight of cement, the sulphate resistance of cement–sand mortar was significantly improved when exposed to sodium sulphate and magnesium sulphate solutions of concentration below 10%. For higher sulphate concentrations, the mortars made with Type V cement and 15% Alberta fly ash developed adequate sulphate resistance. Langan et al. (1983) demonstrated the beneficial effect of Alberta fly ash in improving the sulphate resistance of high volume fly ash concrete containing 50% fly ash as cementitious material.

Fay and Pierce (1989), at the Bureau of Reclamation's concretes and structural laboratories, conducted a systematic study on air entrained fly ash concrete. They selected three fly ashes from the United States, representing a range of CaO contents of 11 to 28.8% and used 10 to 100% fly ash by weight of the total cementitious material, since the fly ashes were also themselves quite cementitious. The specimens were cured for 14 days in a 100% humidity room and for another 14 days in a 50% humidity room before they were immersed in 10% Na_2SO_4 solution or subjected to cyclic soaking and drying phases in 2.1% Na_2SO_4 solution to generate data by accelerated test. The typical results of expansion versus time from the concretes made with different fly ashes at 75% replacement level for two cementitious contents are presented in Figures 6.3–6.6.

The test results show that both low calcium Class C and Class F fly ashes can be effective cement replacements in controlling sulphate expansion. Class F fly ash was the most effective in reducing sulphate expansion at the lower cementitious content, 251.5 kg/m³ (424 lb/yd³). Likewise at the higher cementitious content, 387 kg/m³ (645 lb/yd³), Class F fly ash at 30% replacement level was found to be the most effective. The high calcium Class C fly ash concrete mixes, with fly ash replacement levels below 50%, generally exhibited more expansion than the control mixes, opposite to the behaviour observed with low calcium Class F concrete mixes. For low calcium Class C fly ash replacement levels were suggested to be greater than 30% and for high calcium Class C fly ashes the corresponding suggested levels were greater than 75% to achieve the most improved sulphate durability. The replacement level for a particular type of fly ash was found to depend on cementitious

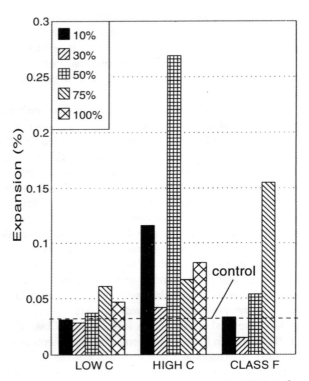

FIGURE 6.3. Expansion of fly ash mixtures with 645 lb/yd^3 of cementitious materials at 1030 days (Soak Test) (Von Fay and Pierce 1989).

material content of the mix. With lower cementitious material content, 50% or more, while for richer mixes 50% or less Class C was suggested for improving the sulphate resistance of concrete. Of course, Class F fly ash at replacement level of 30% significantly improved the sulphate durability for the cementitious levels tested and appeared to be optimum for both test conditions employed in the investigation. The high calcium Class C fly ash concretes generally performed much worse than

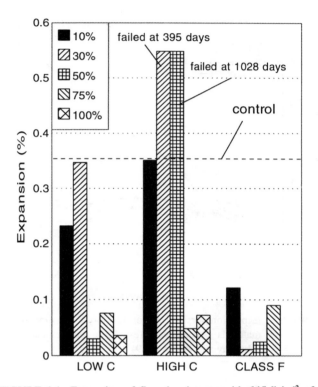

FIGURE 6.4. Expansion of fly ash mixtures with 645 lb/yd³ of cementitious materials at 1030 days (Accelerated Test) (Von Fay and Pierce 1989).

the control mix under sulphate attack indicating the least effectiveness of these types of ash in reducing expansion.

Many researchers have found that alumina content of fly ashes as well as of blast furnace slag has almost similar effect on the sulphate resistance of resulting concrete. With low alumina content, 15 to 16% in fly ash, the concrete mixes having 20% fly ash as cement replacement showed 8% strength gain on immersion in sulphate solution for about 5 months. However,

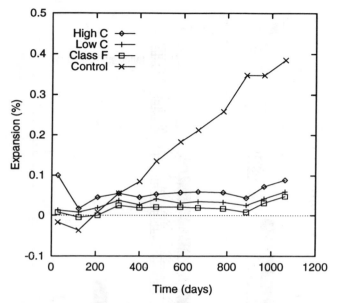

FIGURE 6.5. Expansion versus time for 424 lb/yd³ of cementitious
materials (Accelerated Test; 75% replacement level)
(Von Fay and Pierce 1989).

with the use of fly ash containing about 30% alumina, a strength
loss of 23% was observed under similar conditions of immersion in sulphate solution.

According to Mehta (1993), fly ashes are reported to be
prominent among the group of pozzolans that significantly increase the life expectancy of concrete exposed to chemical
attack, especially sulphate attack. In general, Class F type fly
ash meeting the specification requirements will improve the
sulphate resistance of any concrete/mortar mix in which it is
included, although the degree of improvement may vary with
either the cement used or the fly ash. The situation with Class
C fly ash is different. A few studies indicate that some Class C

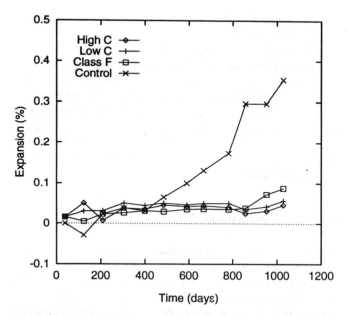

FIGURE 6.6. Expansion versus time for 645 lb/yd³ of cementitious
materials (Accelerated Test; 75% replacement level)
(Von Fay and Pierce 1989).

fly ashes may rather reduce sulphate resistance when used in
normal proportions.

The effect of different types of fly ashes on sulphate durability of concrete will thus usually vary with the characteristics
of the cement as well as the fly ash in addition to their relative
proportions in the concrete mix. It is generally accepted that the
maximum sulphate resistance for the concrete will be achieved
using the most sulphate resistant Portland cement available
along with high quality Class F or low calcium Class C fly ash
at replacement level of about 30% by weight of cement (EPRI
1987). Several construction agencies suggest that in selecting
the fly ash for sulphate durability, one should look for an ash

with the lowest R value and a proven history of satisfactory performance either by laboratory or field tests.

In the literature some researchers have reported the conflicting effects of bituminous and sub-bituminous fly ashes on sulphate resistance of resulting concrete. The detailed characteristics of fly ashes particularly the chemical composition need to be considered to draw definite conclusions. More research need to be conducted to understand the mechanism of sulphate attack on fly ash concrete so that resistance on the basis of an indicator similar to R factor as put forward by Dunstan (1980) can be established.

6.3.2. Sea Water Attack

Relatively few studies are reported in the literature in regard to the effects of fly ash addition in concrete used in marine structures. As is the case with several other aspects of durability, the significant improvement achieved in impermeability with the incorporation of fly ash makes the fly ash concrete potentially favourable construction material under marine environments. The deterioration of concrete exposed to sea water may occur due to the combined physical and chemical actions of sea water on concrete.

It should be noted that ordinary cement concrete in tidal as well as splash zone, subjected to alternate wetting and drying is severely affected, while the concrete below the low tide mark, permanently submerged, is affected the least. The physical actions include cyclic action of sea waves, impact and abrasion due to floating debris and ice, freezing and thawing, temperature gradient, in addition to alternate wetting and drying. The chemical actions involve reactions between the cement/hydration products in the cement mortar matrix and the sulphate, chloride, carbonate and other salt ions abundantly available in sea water. Because of the cumulative damage caused by destructive physical and chemical actions of sea water, the concrete containing fly ash, like normal concrete, will disintegrate and become unserviceable over a period of time.

The problem of chloride induced corrosion of reinforcing steel embedded in marine concrete structures is considerably more severe than in normal land based concrete structures. It is, therefore, essential to provide sufficiently thick cover of high quality concrete made with adequate cement content and low water cement ratio. With the addition of fly ash, calcium hydroxide formed during hydration of cement undergoes reaction with alumina and silica in the of fly ash to form additional calcium silicate hydrates. This results in the prevention of leaching of calcium hydroxide and significant reduction in the permeability of concrete.

In regard to the effects of fly ash addition on resistance of concrete to sea water, the previous discussions on its effect on permeability, sulphate attack, frost, and resistance to corrosion of steel are of great relevance. It is recognized that permeability is the major property affecting various aspects of durability of concrete and this holds equally good for sea water attack. The impermeability of concrete is significantly improved with fly ash and thus fly ash concrete has a potential for use in the marine environment.

Malhotra et al. (1994) have described an ongoing research project since 1978 related to the long term performance of concretes containing mineral admixtures including fly ash under marine environment. The investigations on fly ash concrete specimens were started in 1979 using Class F fly ashes obtained from plants located in Detroit, USA, and in Lingan, Nova Scotia, Canada. The test prisms measuring 305 mm × 305 mm × 915 mm, installed at mid tide level on a rack, are exposed to repeated cycles of wetting and drying, and to about 100 cycles of freezing and thawing per year. The test specimens are planned to be kept at Treat Island, Maine, USA, exposed to the marine environment until the year 2005. The performance of specimens is of course monitored regularly on an annual basis. Malhotra et al. (1994) have reported that after 14 years exposure of the test specimens made with both the normal and semi-light aggregate concretes, containing fly ash or slag or silica fume or a combination of these materials, were in good to excellent condition provided water to cementitious materials ratio is maintained below 0.5 with a certain minimum amount of cement content. They also reported that non-air-entrained concretes totally disintegrated after a few year's exposure at the test site. No significant difference in the performance of concretes made with ASTM Types I, II and V cements was reported.

At present, there is a lack of direct information with regard to the effect of fly ash on concrete as such. Relative impermeability of the fly ash concrete is considered as the most important attribute in enhancing concrete durability under marine environment. A need for research in laboratory as well as in the field is self-evident when investigating the behaviour of fly ash concrete exposed to sea water attack.

6.3.3. Resistance to Alkali–Silica Reaction

This deleterious chemical reaction occurs when certain acidic aggregates, containing reactive silica in the form of silicate, react with alkali metal ions such as sodium and potassium, present in a highly alkaline solution, resulting from the hydration of a high alkali Portland cement. As a result, alkali–silica gel is formed which can cause considerable expansion on water adsorption. The gel is of the unlimited swelling type, it imbibes water with a consequent increase in volume. Since the gel is confined by the surrounding cement paste internal pressures are built up, which eventually lead to expansion, cracking and disruption of the cement paste. It is also believed that it is the swelling of hard reactive aggregate particles that is most harmful to concrete. Some of the relatively soft gel is leached out by water over a period of time and deposited in the cracks already formed due to the swelling of aggregates.

Idorn (1991) has reported that in field cases of deleterious silica reaction, the cement paste is chemically unaffected while the reacting aggregate particles are internally fractured and/or partially dissolved. High alkaline Portland cements generally having more than 0.6% sodium equivalent alkali content is highly vulnerable to attack by aggregates containing reactive amorphous as well as crystalline silica such as chalcedony, crypo-crystalline fibrous, and tridymite. Alkalies in most cases are derived from Portland cement itself, but they can also be augmented by their presence in the mixing water, admixtures, salt contaminated aggregate, and deicing salts used on concrete.

The alkali–silica reaction progresses slowly and the continuous formation of swelling type gel causes internal disruption and cracking of concrete. The damage to concrete due to this reaction is appreciably aggravated when other causes of deterioration such as weathering effects of freeze–thaw, sulphate attack, and other aggressive physical and chemical processes are concurrently active (Mehta 1983).

Stanton (1942) was the first to recognize the deterioration of concrete as a result of reaction between the alkaline hydroxyl ions in the pore water of concrete and certain forms of silica occasionally present in the aggregate including opaline chert and siliceous magnesium lime stone. The studies were conducted using aggregates containing opaline material and cement with acid soluble alkali content of more than 0.6%. Subsequently Stanton (1942) reported that deleterious expansion due to the alkali–silica reaction could be reduced or eliminated by the addition of finely divided mineral admixtures including fly ash containing siliceous material.

In addition to being influenced by the type and characteristics of aggregate, alkali–silica reaction (ASR) or sometimes designated as alkali–aggregate reaction (AAR) is also affected by the available alkali content, presence of moisture, temperature, type of admixture (organic and inorganic) and the alkali to silica ratio of the concrete mix. Numerous early studies suggested the effectiveness of fly ash in inhibiting or reducing expansion resulting from ASR (Stanton 1942, Porter 1964, Pepper and Mather 1959). Figure 6.7 shows a typical plot of expan-

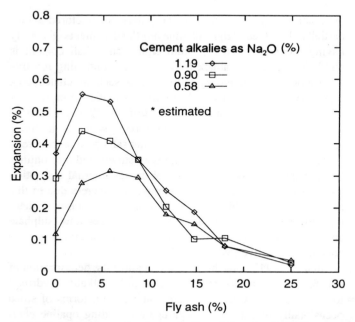

FIGURE 6.7. Effects of cement alkalis and fly ash on alkali–aggregate reaction (Dunstan 1981).

sion versus fly ash content for cements with different equivalent alkali content.

Goldbeck (1956) and Dunstan (1981) identified aggregates and their mineralogical constituents that can react with alkalis in concrete as follows:

- The silica materials: opaline or chalcecondic cherts, tridymite, cristobalite, siliceous limestone.
- Glassy to crypo-crystalline rhyolites, dacites, andesites and their tuffs.
- Zeolite and neulandite.
- Certain phylites.

The reactivity of aggregate is affected by its particle size and porosity as these influence the area over which the reaction can take place. The greater the surface area of the reactive aggregates, the lower the quantity of alkalis available per unit area, and thus the less alkali silica gel can form. Furthermore, because of the low mobility of calcium hydroxide liberated during cement hydration, only that present near the surface of the aggregate is available for reaction, so that the quantity of $Ca(OH)_2$ per unit area of aggregate is independent of the magnitude of the total surface area of the aggregate. With the addition of fly ash, containing pozzolanic silica in fine size fractions, the reactive surface area of the aggregate exposed to given alkaline pore water increases and this results in an increase of calcium hydroxide/alkali ratio of the reactive solution at the boundary of the aggregate. Under such circumstances a nonexpansive calcium alkali silica gel is produced with favourable effects.

Alkalis, particularly Na_2O and K_2O, are highly volatile and are present in high amounts in kiln dust due to combustion process which accompanies cement production (Jawed and Skalny 1977). Current environment and energy concerns, especially initiated by the Resource Recovery and Conservation Act (1976) of the United States, make recycling of particulates removed from the flue gases during cement manufacture, termed as cement kiln dust, an economically attractive procedure. This in turn tends to result in higher alkali cement than have been produced from the same raw material.

The source of alkali is not regarded as important because it is the concentration of soluble alkali in the system that is known to affect expansion due to ASR. Thus soluble alkalis from various ingredients of concrete including admixtures are regarded as equally harmful as that from Portland cement. When a potential for alkali–aggregate reaction exists, it is necessary to evaluate all the ingredients carefully for their alkali content as well as other specified parameters. The relative percentages of fly ash, cement and fine aggregate may be as important as the alkali contents of the fly ash and the cement. The use of Class

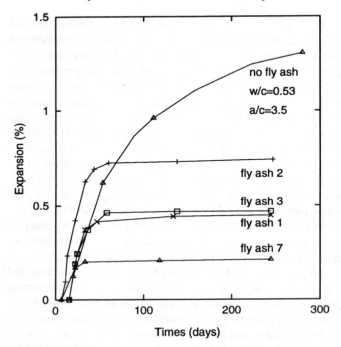

FIGURE 6.8. Variation in expansion with age for specimens with different fly ashes (Hobbs 1981).

F fly ashes in concrete may provide a means by which greater alkalies in the cement can be tolerated for equal or lower unit cost of the cement as well as with a reduction in the amount of cement used. High calcium fly ashes containing large amounts of soluble alkali sulphates, have been reported to increase rather than decrease the rate of damage through alkali–silica reactivity (Mehta 1983).

Hobbs (1981) studied the effect of different Class F fly ashes and blast furnace slag on alkali silica reactivity of mortars made with opaline aggregates and high alkali cement. The results of his study for fly ashes are presented in Figure 6.8 and the

TABLE 6.1. The omposition and properties of fly ash (Hobbs 1981)

Composition	Fly ash number			
	1	2	3	7
SiO_2	50.02	51.48	46.58	49.72
Fe_2O_3	9.02	8.70	14.24	5.22
Al_2O_3	26.83	28.08	25.22	32.45
CaO	1.48	1.27	4.10	2.77
MgO	0.93	0.93	0.95	2.41
SO_3	0.79	1.15	1.29	0.53
Loss on ignition	3.43	1.74	1.84	3.24
Na_2O (total)	0.88	1.13	0.80	0.38
Na_2O (water soluble)	0.07	0.10	0.08	0.02
K_2O (total)	3.90	3.85	2.35	1.40
K_2O (water soluble)	0.07	0.11	0.04	0.02

chemical compositions of the fly ashes used are given in Table 6.1.

Based on the test results, Hobbs (1981) suggested that:

- "The partial replacement of a high-alkali cement by fly ash reduced the long-term expansion due to alkali–silica reactivity but, even when 30% or 40% of the cement was replaced, most of the blended cement mortars cracked at earlier or similar ages as compared to the Portland cement mortars."

- "The effectiveness of the fly ashes in reducing long term expansion varied widely. It is suggested that the effectiveness of the fly ashes may be dependent upon its alkali content or fineness."

- "Where part of the cement was replaced by fly ash, the lowest mortar alkali content, expressed as equivalent Na_2O, at which cracking was observed was 2.85 kg/m^3. This relates only to the acid soluble alkalis contributed by the Portland cement and compares with a figure of 3.5 kg/m^3 for a Portland cement mortar."

- "If it is assumed that fly ash acts effectively like a cement with an alkali content of 0.2% by weight, the lowest alkali content at which cracking was observed was 3.4 kg/m^3."

- "Both fly ash and granulated blast furnace slag act as alkali diluters, slag being more effective in reducing damage due to alkali–silica reactivity than fly ash."

- "From the above it may be concluded that, when the aggregate to be used contains a reactive constituent and when the concrete is to be exposed to external moisture, damage due to alkali–silica reactivity is unlikely to occur if the acid-soluble equivalent Na_2O content of the concrete is below 3 kg/m^3. In calculating the alkali content of the concrete, granulated blast furnace slag may be assumed to contain no available alkalis while fly ash should be assumed to have an available alkali content of 0.2% by weight."

It was further estimated that the minimum alkali content of mortar at which excessive cracking owing to ASR expansion occurred was 3.4 kg/m^3 as acid soluble alkali, equivalent to 2.5 kg/m^3 as water soluble alkali (Hobbs 1981).

The majority of the laboratory studies with regard to ASR have employed very reactive amorphous, opaline aggregates such as Beltan opal from California, and/or pyrex glass. Both are generally more reactive than many of the natural aggregates encountered in practice. While drawing specific conclusions for practical applications on the basis of the test data available in the literature, the above observation should be considered. It may be noted that not all aggregates are vulnerable to ASR, nor do all potentially reactive aggregates behave in the same way.

Similarly, fly ashes from different sources influence ASR significantly differently. Whereas most Class F fly ashes reduce expansion, the use of some Class C fly ashes has been found to be ineffective or deleterious in relations to alkali-reactivity. Current knowledge concerning the role of Class C fly ashes for reducing expansion is not sufficient to draw specific conclusions concerning their overall effectiveness.

In the United Kingdom (Nixon and Gaze 1981) and in South Africa (Oberholster and Westra 1981) hornfels of the Malmesbury group have been examined with regard to alkali–silica reactivity affected by the use of mineral admixtures in mortar/concrete. A study conducted in Canada by Swenson and Gillott (1960) showed that another form of alkali aggregate reaction, alkali–carbonate reaction, takes place between an argillaceous dolomitic limestone aggregate and high alkali cement. The authors also reported that the use of pozzolanic admixtures including fly ash in concrete had no effect in reducing expansion resulting from this type of reaction.

The results of the study by Perry et al. (1987), with 12 different Canadian fly ashes replacing 20 to 40% cement by weight in mortar mixes made with reactive opaline aggregate showed a reduction in expansion after one year ranging from 5 to 81% at 20% replacement level, 34 to 89% at 30% replacement level, and 47 to 92% at 40% replacement level, respectively, compared to that of control mortar mix with no fly ash. The alkali content of fly ashes studied varied over wide limits. However, the alkalies present in fly ashes were less susceptible to reactive aggregate due to their limited solubility in water and being in combined form unlike the free and water soluble alkalies of Portland cement.

Oberholster et al. (1981) studied the effect of adding mineral admixtures, including fly ash, to mortar and concrete made with

Malmesbury group aggregates in order to examine how effective they were in reducing expansion. They observed that fly ash addition effectively suppressed expansion at cement replacement levels of 20% or more on an equal volume basis. Sprung and Adabia (1976) and Hobbs (1982, 1983b) reported that the finer the fly ash, the more effective it was in controlling expansion and damage to concrete through ASR. Their finding was however contradicted by Perry et al. (1987), whose study results indicated that an increase in the surface area of the fly ash was less effective in controlling expansion. Apparently the fly ashes from different sources behave differently in suppressing the ASR.

Presence of moisture is known to be an important factor influencing the ASR. Ludwig (1981) studied the effect of humidity and temperature on mortar bars cured for 3 years and found the critical humidity required to prevent damage through ASR to be less than 85%. After three years, the specimens were transferred to an environment with more than 95% relative humidity (R.H.). The bars cured at 80 and 90% R.H. showed renewed expansion. At an elevated temperature of 40°C, the expansion started earlier but reached lower values at later ages relative to the control bars at 20°C. Thus increasing the temperature at early age increases the rate of chemical reaction occurring between the alkalies and reactive silica in the aggregate and as a result higher expansion takes place during the early periods. The beneficial effect of moderately high temperature on pozzolanic reactivity of fly ash becomes apparent at later ages and thus result in reduced expansion. To study the effect of temperature, Diamond et al. (1981) prepared mortar bars stored at 20 and 40°C after 24 hours of casting. Exposure to 40°C for 2 days resulted in expansion higher than that at 20°C for 30 days.

The results of a recent study by Joshi et al. (1994) to investigate the effect of Alberta fly ashes used at replacement levels of 20 to 60% on alkali–aggregate reactivity of mortar mixes made with high alkali cement and opal aggregate are presented

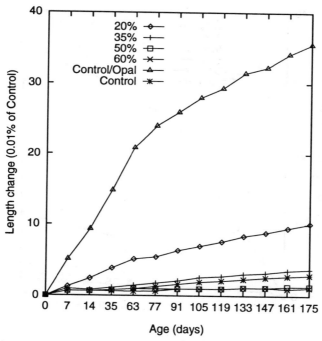

FIGURE 6.9. Length change of mortar bars with 20, 35, 50 and 60%
Wabamun fly ash as replacement of cement (Joshi et al.
1994).

in Figures 6.9 and 6.10. Effective suppression of expansion due
to ASR was achieved with the incorporation of Alberta fly ashes
at replacement levels of 35% and higher. The control mixes
made with opal as aggregate used at pessimum proportion exhibited much more enhanced expansion than the corresponding
mixes with natural aggregates. The authors observed favourable
effects of fly ash on the alkali–aggregate reactivity in concrete
mixes containing up to 60% fly ash as cement replacement.

The expansion owing to ASR in concrete results in a characteristic form of cracking often referred to as "pattern" or map

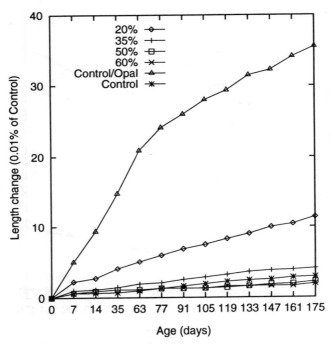

FIGURE 6.10. Length change of mortar bars with 20, 35, 50 and 60%
Sundance fly ash as replacement of cement (Joshi et
al. 1994).

cracking (Gillott 1980). It was reported that for a specific cement of a given alkali content, the maximum expansion occurs
at some particular proportion of reactive aggregate to total aggregate. This particular proportion is termed as the "Pessimum
Proportion." This phenomena is governed by the ratio of alkali
in cement to silica in the aggregate (Strubble and Diamond
1986).

When a part of the Portland cement is replaced by fly ash,
the general perception that reduction in expansion caused by
ASR is simply due to the reduction in alkali content, which may

result from cement replacement is justified, only when the fly ash itself does not contain water soluble alkali. Several studies have refuted this simple explanation, because in a study by Gaze and Nixon (1983) the alkali content of some fly ashes was higher than that of the cement replaced. An alternate and more scientific explanation for reduction in expansion with fly ash addition might be given in terms of physicochemical processes including hydration reaction of cement and pozzolanic reaction of fly ash. Fly ashes as well as granulated blast furnace slag act as alkali-diluters, slag being more effective than fly ash in reducing damage due to ASR. It is postulated that the reaction between the very small particles of amorphous silica glass in the fly ash and the alkalies in the Portland cement and the fly ash, ties up the alkalies in an innocuous manner by producing nonexpanding calcium–alkali–silica gel. Thus hydroxyl ions remaining in the solution are insufficient to react with the material in the interior of the larger reactive aggregate particles and thus disruptive osmotic forces are not generated. Because of fine particle size, fly ash addition not only improves the packing of cementitious material but also reduces the permeability of cement paste due to the formation of new hydration products by the pozzolanic reactions. The ion migration and availability of moisture needed for ASR are considerably reduced, thereby significantly improving the resistance of concrete to the deleterious action of ASR.

The large size of ions of harmful elements, such as Na from deicing salts, can not easily penetrate the hardened fly ash concrete because of decreased pore size and permeability. Therefore, alkali–aggregate reactions are retarded or minimized by use of fly ash in concrete.

The choice of low alkali cement has traditionally served to avoid disruptive expansions with aggregates susceptible to ASR. Since a number of fly ashes, particularly high calcium Class C ashes, may have appreciable amounts of soluble alkalies, there is a danger of increased alkali–silica reactions under some circumstances. Some studies report the use of high levels of fly ash to suppress the ASR expansion effectively. For such situations the reproportioning of mixes is suggested along with the use of superplasticizer and air entraining agent to develop the specified strength at early ages. Most researchers emphasize the need to conduct tests on the ingredients to be used in the field proportions to ensure that expansion will be reduced to safe levels in the long run.

Since 1974, seven international congresses on alkali–silica reaction have been held and numerous research studies have been presented. Despite the continuous and sustained world-wide research over the last 50 years, the mechanism and details of control of expansion caused by ASR are not fully understood. Further research studies need to be conducted for developing a better understanding of the problematic phenomenon so that it can be effectively and economically controlled and prevented.

6.4. CARBONATION OF CONCRETE

In recent years, the effect of fly ash on carbonation of concrete has gained some attention, because this process is believed to initiate and aid the chloride induced corrosion of reinforcing steel bars embedded in concrete.

Carbon dioxide (CO_2) in the atmosphere reacts, in the presence of moisture, with calcium hydroxide ($Ca(OH)_2$) to form calcium carbonate ($CaCO_3$). Similarly the hydrated cement compounds also react with CO_2, although to a lesser extent. The effects of carbonation have been observed even at the low partial pressure of 3×10^{-4} atmosphere of CO_2 in normal atmosphere. The mechanism of carbonation is attributed to solution of gas in the pore fluid which forms carbonic acid (H_2CO_3) and the diffusion of gas through preexisting microcracks in concrete and subsequent solution and reaction.

The result of laboratory studies demonstrate that carbonation occurs in all types of concrete and the rate of this process depends upon several factors such as permeability and size of the specimen, ambient temperature and humidity conditions, mix proportions, moisture content, and amount of $Ca(OH)_2$ available for reaction. The reported possible deleterious effects of carbonation in concrete are increase in permeability, increased shrinkage and cracking, and increase in corrosion potential of embedded steel reinforcement.

Until recently, very few studies were undertaken on the carbonation of fly ash concrete owing to the limited or rather restricted use of fly ash concrete in reinforced and prestressed concrete structures. Kasai et al. (1983) studied the carbonation of mortar specimens made with different types of cement and fly ash after 7 days of moist curing. The carbonation was observed to progress rapidly up to three months and after that it slowed down. As shown in Figure 6.11 the greater the coefficient of permeability of the specimen, the greater was its susceptibility to CO_2 attack manifested in terms of increased depth of carbonation. The specimens made with fly ash cement exhibited greater carbonation effect than ordinary Portland cement specimens.

Ho and Lewis (1983) tested three types of concrete mixes, one using Portland cement without any additives, the second containing a water reducing admixture, and the third in which fly ash was used to replace part of the cement. The object of the study was to determine the extent of carbonation after subjecting the specimens to 4% CO_2 at 20°C and 50% R.H. for 8 weeks. The results of the study suggested that with 7 days of curing in CO_2 atmosphere, the fly ash concrete specimens tended to have a greater depth of carbonation, but as the curing period was extended to 90 days, the fly ash concrete showed improved resistance to carbonation as shown in Figure 6.12.

Nagataki et al. (1986) reported a direct relationship between compressive strength at 28 days and depth of carbonation irrespective of fly ash replacement in concrete. As can be seen in Figure 6.13, the extent of carbonation decreased with an increase in compressive strength.

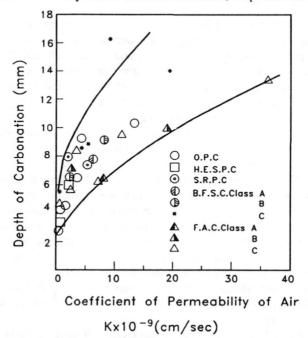

FIGURE 6.11. Relationship between coefficient of permeability of air and depth of carbonation at 6 months (Kasai et al. 1983).

    Gebauer (1982) from a similar study observed that an increase in water cement ratio of concrete mix resulted in an increase in the depth of carbonation. Furthermore, in an investigation on carbonation behaviour of cement sand mortars made with 25% pulverized fuel ash or fly ash replacing cement and at water cement ratios of 0.35 to 0.55, Butler et al. (1983) reported that more carbonation occurred at lower water contents when the fly ash concrete was desiccated with calcium chloride ($CaCl_2$). However, in the case of nondesiccated specimens, the effect of fly ash on carbonation at lower water contents was

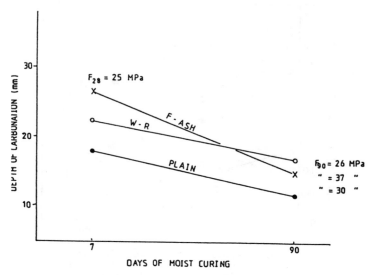

FIGURE 6.12. Influence of curing duration on depth of carbonation (Ho and Lewis 1983).

insignificant. Of course in nondesiccated specimens, more carbonation was observed at high water cement ratios.

In a recent study, Joshi et al. (1994) found that up to about 7 days, the extent of carbonation measured by the affected depth from the outer surface in concrete, after subjecting the specimen to 4% CO_2 at 20°C and 50% R.H., was more in concrete containing fly ash than the control concrete without fly ash. However, after 90 days curing, the trend reversed in that the fly ash concrete exhibited less carbonation than the control concrete. See data in Table 6.2.

Much of the recent experimental work on carbonation of fly ash concrete involved exposure of relatively immature concrete to 50% R.H. and 4% CO_2 atmosphere which is at least five times that of exposure under normal field conditions. Fly ash

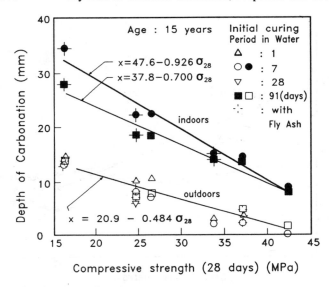

FIGURE 6.13. Relationship between compressive strength and depth
of carbonation (Nagataki et al. 1986).

concrete under field conditions will rarely be exposed to that
high CO_2 concentrations at early ages and at low R.H. 50%,
because longer moist curing periods which allow pozzolanic
reactions in fly ash concrete should reduce the rate of carbonation. In this respect, the results of Tsukyama et al. (1980) on fly
ash concretes exposed for long periods in the field conditions
in Japan are of interest. They reported that fly ash concrete with
identical water to cement ratios exhibited reduced depth of
carbonation or termed neutralization by the authors. Also,
longer periods of moist curing under outdoor exposure resulted
in reduced depth of carbonation for all the mixes.

There are substantial published data to show that good quality fly ash concrete properly proportioned, well compacted and
adequately cured to develop pozzolanic reactions will have

TABLE 6.2   Depth of carbonation and splitting tensile
strength of different concrete mixes

Mix no.	Fly ash replacement level (%)	4% CO_2 (8 weeks)		100% CO_2 (1 day)	
		Depth (mm)	Tensile strength (MPa)	Depth (mm)	Tensile strength (MPa)
FS40-I	40(S)	0.38	3.7	0.18	4.2
FS60-I	60(S)	0.85	3.7	1.46	3.9
FF40-I	40(F)	0.70	2.9	0.43	2.6
FF60-I	60(F)	3.59	3.0	10.33	2.6
FW40-I	40(W)	0.87	3.4	0.36	2.9
FW60-I	60(W)	9.53	1.9	6.06	1.9
C-I	0	0.15	4.2	0.08	3.8
C-III	0	0.18	3.7	0.20	3.7

I = ASTM Type I Cement. III = ASTM Type III Cement. S = Sundance, F =
Forestburg, W = Wabamun.

equal or sometimes better resistance to carbonation than corresponding plain concrete. Essentially, the effects of fly ash addition on resistance to carbonation follow the same trend as on
strength and permeability of concrete, as a result of physicochemical processes associated with hydration and pozzolanic
reactions of fly ash–cement–water system. The addition of fly
ash due to its pozzolanic and sometimes also cementitious reactions not only consumes the free lime present in the cement
paste, but also decreases the water permeability and ion dif-
fusivity of the system, thus improving the overall resistance of
concrete to carbonation. Because of the slow rate of pozzolanic
reactions, the beneficial effects of fly ash addition, however,
become apparent after long periods of curing. It is thus believed
that the apprehension of carbonation of fly ash concrete to
adversely affect the corrosion of reinforcing steel bars may not
be realistic and justified.

6.5. RESISTANCE TO CORROSION OF REINFORCING STEEL IN CONCRETE

In recent years, the problem of corrosion of reinforcing steel embedded in fly ash concrete has become of great concern for applications of concrete in the construction of reinforced and prestressed concrete structures, particularly, those subject to chloride induced corrosion of steel caused by the use of deicing salts or sea water exposure of structures in a marine environment. The bond between the surrounding concrete and steel reinforcement and the high alkalinity of the concrete provide protection from corrosion.

In a hydrated Portland cement paste, about 20% $Ca(OH)_2$ by weight of the hydration products is present to provide the reserve basicity for steel protection. Diamond (1981) has reported that the high alkalinity of the pore solution in cement paste primarily results from the presence of sodium and potassium ions rather than from the presence of calcium hydroxide alone. In the two fly ash systems examined, Diamond reported that pH of pore solution was reduced from 13.75 in a control system to about 13.55 in the presence of fly ash. Under highly alkaline conditions i.e. pH lager than 11.5 of pore solution in concrete, a protective iron oxide film forms on the surface of reinforcing steel that makes it passive against further corrosion. This passive iron oxide layer is susceptible to the attack by chloride ions in concrete or to carbonation when the pH of the surrounding concrete is reduced below 11.0. Once this passive layer is destroyed, a galvanic cell can form between different areas on reinforcing bars causing reduction at anodic area. Furthermore, the rate and extent of corrosion of embedded steel depends on the electrical conductivity of the surrounding concrete and the permeation of moisture and air through the concrete. In practice, adequate cover of high quality and impervious concrete over the reinforcing steel has been found to provide adequate protection against corrosion.

Several studies have shown that properly proportioned and cured fly ash concrete, when used as surrounding medium for the reinforcing steel, the corrosion protection remains unaffected compared with that of normal concrete (Kondo et al. 1981, and Paprocki 1970). Andrade (1986) tested concrete mixes with and without fly ash for corrosion using polarization resistance techniques where the intensity of corrosion of the steel reinforcement was measured and compared to safe values of corrosion. The addition of fly ash promoted the corrosion of steel in mortars but had no effect on concrete specimens. The decrease in the alkalinity due to introduction of fly ash was reported to have a major effect in promoting corrosion in fly ash mortar mixes. Larsen et al. (1976) in their study found an improvement of corrosion protection of steel with the inclusion of fly ash in concrete.

The incorporation of fly ash in concrete may influence alkalinity, permeability, carbonation, and also electrical conductivity of the concrete cover surrounding the steel. In light of the comments of the Rilem Technical Committee on corrosion of steel in concrete (1974), the effects of fly ash on corrosion resistance of concrete should be contingent on the physicochemical changes in concrete which result with its addition. Some of the important observations based on previous research studies are described below.

Carbon and sulphur compounds present in fly ash are normally limited by specifications and they are not materially different from those present in concrete whether fly ash is used or not. For example, if carbon in fly ash in below 3%, its percentage in the concrete becomes so small that if it is well dispersed, its effect on electrical conductivity of concrete and, therefore the corrosion of concrete is insignificant. Most sulphur in fly ash is present as sulphate and therefore would have an effect similar to the sulphur components in Portland cement.

Schiepl and Raupack (1992) studied the electrolytic resistance of fly ash concrete made with 20% cement replaced with fly ash compared with that of other concretes. They found that after one month, the fly ash concrete had equal electrolytic resistance to that of plain concrete. After one year, the fly ash concrete exhibited values of the same order of magnitude as blast furnace slag cement concrete containing 74 wt% slag.

It has been stressed in previous sections that due to pozzo-lanic reaction between lime and fly ash, permeability of concrete is considerably reduced thereby decreasing ingress of moisture, oxygen, carbon dioxide, chlorides, and other aggressive salt solutions. Even though the pozzolanic reactions reduce the amount of calcium hydroxide present, adequate alkalinity (pH > 11.5) remains to preserve the passivity of the steel necessary to prevent corrosion (Diamond 1981). It has been reported that sufficient alkalinity remains to preserve passivity at as high as 75% replacement of the cement in concrete. In certain cases, the inclusion of fly ash may rather improve the corrosion protection of steel in concrete when properly proportioned and adequately cured for gaining proper maturity. If fly ash concrete is made with poor quality fly ash, inadequately proportioned and cured, it becomes highly permeable particularly at early ages. Under such conditions, chlorides as well as moisture and oxygen penetrate the concrete and the protective layer at the reinforcing steel surface can be destroyed.

As the steel corrodes, it expands and develops stresses which eventually induce cracking, delamination and spalling of the concrete.

6.6. RESISTANCE TO ELEVATED TEMPERATURES

Recent developments in the application of concrete in structures required to withstand elevated temperatures under some circumstances have led to the necessity of investigating the behaviour of fly ash and other concretes exposed to high temperatures. The use of concrete in structures such as prestressed concrete pressure vessel reactors and other nuclear reactor containment structures, water desalination concrete tanks, launching pads of space vehicles and runways for high speed jet aircraft, and concrete floor under boilers and chimneys, represents some of the possible situations for high temperature exposure of concrete.

It is known that the behaviour of concrete at elevated temperatures is governed by several factors such as duration of temperature exposure, thermal cycling, shock and gradient, concrete composition, moisture content of concrete, pore structure of the cement paste, and the size of concrete member. The deterioration of concrete at high temperature may appear in the form of cracking, strength reduction and sometimes excessive spalling, especially when the moisture content of concrete is very high.

Studies by Nasser and Lohtia (1971), Nasser and Marzouk (1979), and Carette et al. (1982) and by many other researchers have revealed that two distinct situations need to be considered when concrete is subjected to elevated temperatures. In one case, the heating is done under open conditions with the contained moisture in concrete free to evaporate, while in the other case, the heating is done under sealed conditions with the contained moisture prevented from escaping. In the latter case, as is in mass concrete structures for nuclear reactor containment the interior concrete during heating at high temperature will have its moisture converted into steam and thus subjected to steam pressure in addition to other forces. Loss of strength and changes in other structural properties of concrete are much more marked in such situation than heating in open atmosphere.

Water is present in hardened concrete in several forms such as capillary water as free water, gel water or zeolitic water as adsorbed water and chemically combined water as nonevaporable water. Schnieder (1982) has reported that free water is lost when the concrete is heated to 100°C and that adsorbed water starts to evaporate around 180°C. Water in the calcium hydroxide begins to dehydrate around 500°C and finally chemically combined water in hydrated compounds CSH and CAH gets dislodged and starts to escape around 700°C. The deterioration of concrete at high temperature is believed, in general, to be caused by the loss in strength of cement paste due to the decomposition of cementing compounds and high tensile stresses caused by the restraint of paste offered by aggregate.

The test data in regard to high temperature effects on concretes made with normal Portland cement, slag and fly ash at sustained temperatures up to 600°C from the study by Carette et al. (1982) are presented in Figure 6.14. Most previous studies indicate that fly ash addition does not appear to influence the behaviour of concrete at high temperatures. The extent of damage as a result of high temperature exposure is about the same for both fly ash concrete and normal Portland cement concrete. At moderately high temperatures fly ash concrete shows favourable gain in strength due to increased pozzolanic reactivity. However, at high temperatures, loss of strength and changes in other structural properties are observed to be about the same for both plain and fly ash concrete.

6.7. ABRASION/EROSION RESISTANCE

The degradation of normal and fly ash concretes by several other external physical causes — such as abrasive and erosive forces due to moving traffic, flowing water, floating debris and ice, wave action on water front marine structures, and cavitation pressure on the downstream side of the spillway — is of interest to the engineers. The use of hard aggregates and low water to cement ratio has been found to be quite effective in increasing abrasion/erosion resistance of both types of concrete, i.e. with and without fly ash. With the use of fly ash in concrete the quality of cement paste is improved, and the leachability of calcium hydroxide is impeded with age. Because of the dense structure of cement paste and good bonding characteristics of fly ash concrete, it is believed that the ten-

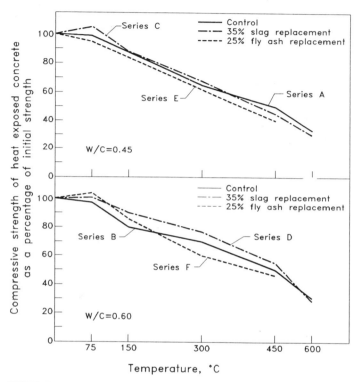

FIGURE 6.14. Compressive strength of concretes after 1 month of
exposure to various elevated temperatures (Carette et
al. 1982).

dency of coarse aggregate to be plucked out of the binding
matrix by abrasive action is reduced. At equal compressive
strengths, properly cured and finished concrete with and without fly ash will exhibit essentially equal resistance to abrasive–
erosive forces. Use of fly ash thus affects this aspect of durability only to the extent that it usually improves compressive
strength of concrete due to its pozzolanic activity with time.

CHAPTER 7

Applications of Fly Ash in Special Concretes

7.1. ROLLER COMPACTED CONCRETE

During the last two decades, a rapid construction method using no slump concrete and employing the conventional techniques of geotechnical engineering has been widely used for the construction of massive and/or flat concrete placements having large construction areas. The method is termed roller compacted concrete, abbreviated as RCC and involves the depositing of no slump, i.e., dry concrete, in continuous horizontal lifts which are then consolidated by vibrating rollers. In a way the method is more like the procedures used in the construction of earth dams, embankments, road bases, and other earth structures using natural and stabilized soils unlike conventional casting of concrete in forms.

Essentially RCC involves the same types of constituents as ordinary concrete except for the cementitious material components. In RCC, the cementitious material contains significantly larger quantities of pozzolanic materials such as blast furnace slag, and fly ash than in conventional concrete. However, the cementitious material content in RCC is of the order of 100–270 kg/m^3 which is much smaller than in ordinary concrete. Fly ash at the level of 60 to 80% of the cementitious material has been successfully used in several RCC structures in many countries both for technical and economical considerations.

Variations of RCC have been used in pavement construction since the 1960s. Lime and cement stabilized soils and mixtures of soil and aggregates have been used for pavement construction since the 1940s. Since the broad use of RCC as an alternative to regular concrete in the 1970s, it has been successfully used in constructing a wide range of structures such as gravity dams, foundations of large span bridges and power houses, container handling facilities, dry land log sorting areas, composting facilities, and road construction in many countries all over the world (Cannon 1972). In the United States, RCC was used in the construction of Upper Still Water Dam, largest concrete gravity dam for water storage, replacing 75% of cement with fly ash (Dolen 1990a,b).

The four major advantages from the high volume usage of fly ash in RCC are:

(1) higher density due to larger volume of fine particles,

(2) reduced temperature rise due to the slower heat evolution because of lower amount of cement and pozzolanic reactions,

(3) increased long term strength,

(4) reduced cost of construction materials.

In the construction of Upper Still Water Dam, over 225,000 tonnes of fly ash were used. This was one of the largest single applications of fly ash in concrete in the world. Both the structural as well as RCC concretes contained fly ash ranging from about 23 to 70% by weight of cementitious materials, respectively. For the main body of the dam, no slump concrete was mixed in conventional drum or shaft driven paddle mixers, transported by rear end discharge dump trucks, spread into horizontal layers, 0.3 to 0.9 m thick, by a bull dozer, and compacted by a smooth drum vibratory roller. Compressive strength development of fly ash concrete depended on fly ash content, the water to cementitious material ratio, and the chemical composition of cementitious materials. Water reducing and air entraining agents were also used as admixtures to control water to cementitious material ratio and air content for improving freeze–thaw durability. Compressive strength of RCC made with 70% fly ash as cementitious material was in the range of 12.7 MPa at 28 days to over 51.0 MPa at 365 days. Permeability of the RCC was reported to be equal to, or less than, conventional mass concrete.

The use of high fly ash contents in RCC mixes significantly reduced the heat of hydration and thereby rise in temperatures of the concrete. The initial setting time and heat rise of high fly ash RCC mixes were retarded by low concrete temperatures, sulphates in the fly ashes, and by water reducing admixtures. This led to the increased bond potential between different placement layers or lifts.

A comprehensive report for practical use of RCC was brought out by the American Concrete Institute (ACI) Committee 207 in 1980. It was emphasized that the primary role of fly ash in RCC is to occupy space between larger particles because of its finely divided nature. More fly ash content in cementitious paste will provide larger volumes of fine material that would otherwise require the use of additional cement to obtain maximum density of the compacted mass. It is reported that pozzolanic reactivity of fly ash is not of paramount importance, as compressive strength of RCC can develop significantly over long periods of time after placements. Thus, even substandard fly ash, as well as other minerallic materials, in finely divided form, can be suitable for most roller compaction applications.

Papayianni (1992) reported the successful use of a local fly ash, rich in lime and sulphates, and not meeting the existing specification requirements, in the construction of a large RCC dam in Northern Greece. The RCC mixes containing 80% fly ash as cementitious material component developed adequate strength and elastic properties and performed satisfactorily in the structure.

RCC is in fact fly ash modified lean concrete containing high volume of fly ash and very low water content. RCC has been extensively used for road bases and pavement construction in Europe, Japan, North America and many other countries (Sherwood and Potter 1982) in addition to its use in dam construction.

Joshi and Natt (1983) studied geotechnical properties of cement/lime fly ash mixes for use in RCC. They tested a total of 28 mixes, 12 containing cement and fly ash, 12 lime and fly ash, and 4 cement, lime and fly ash as cementitious materials. In the test mixes, the cement/or lime content varied from 1.8 to 4.5% of the total mix material. Amount of fly ash ranged from 17.8 to 35.6% and aggregate content varied from 60 to 80% of the total mix. Fly ash comprised about 80% of the total cementitious material. Based on the results, the authors reported that roller compactable mixes, termed GEOCRETE, can be produced from cement/or lime, fly ash, sand and gravel in a similar way as stabilized soils for embankment and road construction. Mixes containing as little as 20% of 1:8 cement–fly ash as cementitious material developed compressive strengths of 5.5 MPa and 15 MPa at the ages of 7 days and 60 days respectively. The mixes exhibited good durability characteristics as evaluated from drying shrinkage and freeze–thaw tests.

ACI Committee (1980) describes RCC as dry or no slump concrete that is placed in small lifts by using paver or bull dozer and compacted by using vibratory rollers. Typical mix proportions and corresponding strength data of RCC mixes as reported by ACI (1980) are given in the Tables 7.1 and 7.2.

The Committee has reported that presence of both air entraining and water reducing agents in RCC may effectively reduce the vibration time for full consolidation of the concrete. Recently, RCC mix was designed for waste recycling and waste composting facility in the United States to obtain 28 day compressive strength of 27.6 MPa (PCA 1993). The batch quantities per cubic meter of mix were Type I cement = 152 kg (335 lb), Class F fly ash = 87 kg (192 lb) fine aggregate = 653 kg (1440

TABLE 7.1 Mix proportions for some roller-compacted concrete (ACI Committee 207 1980)

Source	Max. agg. size, cm	Mix data, kg/m³				
		Cement	Pozzolan	Water	Fine agg.	Coarse agg.
1	7.6	56	77	77	60	1649
2	11.4	139	0	80	618	1774
3	7.6	139	0	86	683	1691
4	7.6	139	0	83	676	1602
5	7.6	42	78	83	676	1602
6	3.8	75	164	89	745	1426
7	3.8	45	178	84	727	1438
8	3.8	116	139	103	657	1438

TABLE 7.2 Properties of some roller-compacted concretes (ACI Committee 207 1980)

Source	Age, days	Compressive strength, MPa	Shear strength, MPa	
			Mass	Joint
1	138	23	4	2
2	72	26	5	1
3	66	23	6	–
4	120	23	6	3
5	120	16	4	1
6	90	18	–	3
7	90	41	2	2
8	90	–	–	–

lb), coarse aggregate= 907 kg (2000 lb), water = 85.3 kg (188 lb), water reducer = 0.9 kg (32 oz.). The RCC mix was placed using an asphalt paver to the required thickness of about 19 cm (7.5 in) and was compacted by a steel drum vibratory roller.

RCC thus provides a promising option for high volume utilization of fly ash in various civil engineering applications including dam rehabilitation, liners and heavy duty pavements.

7.2. HIGH VOLUME FLY ASH STRUCTURAL CONCRETE

The proportion of fly ash used as a cementitious component in concrete depends upon several factors. The design strength and workability of the concrete, water demand and relative cost of fly ash compared to cement are particularly important in mix proportioning of concrete. From the review of literature, it is generally found that the optimum proportion of fly ash in the cementitious material for high strength (40 MPa), high workability (100 mm slump) concrete is in the range of 30 to 40%; medium strength (20 to 30 MPa), medium workability (50 to 75 mm slump) concrete 50 to 65%; and low strength (10 to 20 MPa), low workability (zero slump) concrete 70% to 90%. As explained in Section 7.1, higher proportions of fly ash, in the range of 70 to 80%, have been successfully used in roller compacted concrete mixes.

Recent studies at CANMET and University of Calgary have indicated that structural concrete with 28 days strength around 60 MPa and of adequate durability can be produced with Canadian fly ashes replacing up to 60% cement by weight and by incorporating high range water reducer and air entraining admixtures in concrete. A number of high volume fly ash concrete (HFCC) field applications have since been undertaken in the United Kingdom, Europe and North America, although most research on HFCC has been conducted in laboratory till recently (Dunstan et al. 1992).

Tarun et al. (1989) presented two case histories involving application of HFCC in the field. In one case, a concrete mix with 70% of cement replaced by Class C fly ash was used to pave a 254 mm thick roadway. To obtain high workability and durability, a water reducer and an air entraining agent were added to the concrete mix. The air content of the concrete was maintained between 5 and 6%.

The main problem encountered with use of HFCC was that the rate of compressive strength gain was slower than anticipated. Otherwise, the condition of the roadway surface was excellent with no defects observed. The other case reported by the same authors involved placing of the same HFCC in the construction of 138 kV transformer foundation measuring 4.9 m × 5.5 m × 1.5 m. No problems were reported during or after construction in both projects and the use of HFCC resulted in considerable economy and technical benefits.

Hague et al. (1984b) conducted tests on concrete mixes in which bituminous fly ashes formed up to 75% of the cementitious material content which varied between 325 and 400 kg/m³. They concluded that such high fly ash concretes with adequate strength, volume stability and durability have a great potential for use in structural applications particularly in pavements. Gifford et al. (1987) showed that concrete with high quality Alberta fly ash replacing cement by 40% can be successfully used in curb and gutter construction in cold climate. Results of their study showed that with adequate field curing the strength of fly ash concrete curb and gutter can be equal to or better than that of plain concrete. With air content of around 6%, the fly ash concrete exhibited freeze–thaw durability similar to the control plain concrete mix.

Langley et al. (1989) reported two case histories where HFCC was used with Class F fly ash constituting 55% of the cementitious material along with a superplasticizer. The mix proportions of one cubic metre of HFCC included Portland cement (Type I) = 180 kg, fly ash (Class F) = 220 kg, coarse aggregate (19 mm maximum) = 1100 kg, fine aggregate (natural sand) = 800 kg, water = 110 kg, and superplasticizer (naphthalene based) = 6 litres. In one case, where columns, beams and floor slab in a building complex required 50 MPa concrete at 120 days, the HFCC use yielded concrete with 74 MPa compressive strength at 120 days, thus exceeding the strength requirement. No unexpected problems were reported and the HFCC proved to be an economical solution for the particular project.

In the second case, the same type of HFCC with 14 mm size maximum aggregate was used in drilled caisson piles larger than 1.0 m in diameter with average length of about 21 m and socketed into bed rock for a depth of 1.8 m. The large diameter piles were to support a 22 storey office lower on the Halifax Water Front, Nova Scotia in Canada. Since the piles were heavily reinforced, the concrete was superplasticized to increase the workability. The minimum 28 day compressive strength requirement of 45 MPa for the pile concrete was easily met by the high fly ash concrete which had compressive strengths of the order of 32 MPa and 51 MPa at 7 days and 28 days, respectively. Additionally, in-situ load tests revealed that bond strength developed between the concrete and rock was quite adequate, with test values exceeding 3 MPa, about 2.5 times of the design requirement of 1.2 MPa.

Sivasundaram et al. (1990) studied in detail the properties of concretes with a wide range of Canadian fly ashes at 58% of the total cementitious materials. The physical properties and chemical composition of cement and fly ashes are given in Table 7.3. The details of mixes along with properties in fresh state age presented in Table 7.4.

The superplasticized and air entrained concretes were tested for compressive strength, creep strain and resistance to chloride ion penetration at various ages up to 1 year. The authors reported the following conclusions.

TABLE 7.3 Physical properties and chemical analyses of cement and fly ashes (Sivasundaram et al. 1990)

	ASTM Type I Portland cement	Fly ash 1	Fly ash 2	Fly ash 3	Fly ash 4	Fly ash 5	Fly ash 6	Fly ash 7
Physical Tests								
Fineness, retained on 45 μm, %								
- wet sieving	–	17.3	19.2	21.2	33.2	19.4	46.0	–
- dry sieving (Alpine Jet)	–	12.3	14.0	16.1	26.4	14.3	33.0	16.3
Blaine specific surface, m²/kg	417	289	198	448	215	326	240	3.7
Autoclave expansion	0.16	–	–	–	–	–	–	–
Specific gravity	3.15	2.56	2.96	2.38	1.90	2.05	2.11	2.54
Compressive Strength, MPa								
3-day	24.7	–	–	–	–	–	–	–
7-day	30.5	–	–	–	–	–	–	–
28-day	38.7	–	–	–	–	–	–	–
Coal type	–	Bit.	Bit.	Bit.	Sub-b.	Sub-b.	Sub-b.	Bit.
Chemical Analyses								
SiO_2	21.5	47.1	38.3	45.1	55.7	55.6	62.1	48.2
Al_2O_3	4.0	23.0	12.8	22.2	20.4	23.1	21.4	24.9
Fe_2O_3	2.56	20.4	39.7	15.7	4.61	3.48	2.99	18.9
CaO	62.7	1.21	4.49	3.77	10.7	12.3	11.0	2.8
MgO	3.70	1.17	0.43	0.91	1.53	1.21	1.76	1.10
Na_2O	0.48	0.54	0.14	0.58	4.65	1.67	0.30	0.59
K_2O	0.67	3.16	1.54	1.52	1.00	0.50	0.72	1.87
TiO_2	0.21	0.85	0.59	0.98	0.43	0.64	0.65	–
P_2O_5	0.07	0.16	1.54	0.32	0.41	0.13	0.10	–
MnO	–	0.78	0.20	0.32	0.50	0.56	0.69	–
BaO	–	0.07	0.04	0.12	0.75	0.47	0.33	–
SrO	0.06	–	–	–	–	–	–	–
SO_3	3.9	0.67	1.34	1.40	0.38	0.30	0.16	0.78
Loss on ignition	1.42	2.88	0.88	9.72	0.44	0.29	0.70	3.7

TABLE 7.4 Mixture proportions and properties of fresh concrete (Sivasundaram et al. 1990)

Fly ash source	Batch	Mixture proportions						Properties of fresh concrete			
		water to binder ratio	agg./cement	cement kg/m³	fly ash kg/m³	super-plasticizer	air-entraining admixture ml/m³	unit weight kg/m³	temperature °C	slump mm	air content %
1	A	0.22	7.8	228	315	15.9	978	2445	21	200	3.3
	B	0.21	7.8	229	316	15.8	735	2455	18	200	2.8
	C	0.31	12.3	154	211	4.3	239	2375	22	140	5.0
	D	0.31	12.4	151	208	4.7	347	2345	21	100	5.7
2	A	0.22	8.0	224	310	14.5	800	2450	18	200	4.0
	B	0.22	8.0	219	301	14.2	781	2390	20	200	5.4
	C	0.31	12.6	149	205	3.3	310	2340	21	190	6.6
	D	0.30	12.6	150	206	3.3	311	2350	20	125	7.0
3	C	0.32	12.3	154	211	14.5	719	2375	23	200	3.5
	D	0.32	12.3	155	212	14.6	884	2390	24	200	3.5
4	A	0.22	7.3	224	309	21.5	1596	2295	19	200	4.6
	B	0.23	7.3	222	306	21.9	1583	2275	20	200	4.7
	C	0.31	11.9	153	210	5.4	429	2295	22	200	5.0
	D	0.31	11.9	153	211	5.4	430	2300	21	200	5.0
5	A	0.22	7.4	223	307	22.7	1909	2320	22	200	4.8
	B	0.22	7.4	222	306	22.7	1902	2310	21	225	4.7
	C	0.31	12.1	151	208	4.6	551	2295	21	175	6.4
	D	0.31	12.0	154	211	4.3	447	2325	22	140	5.3
6	A	0.25	7.5	212	292	41.2	1513	2235	21	200	6.8
	B	0.25	7.5	215	296	41.6	1530	2260	20	200	5.8
	C	0.32	12.1	152	208	11.2	284	2315	22	200	4.9
	D	0.32	12.1	148	204	11.2	277	2260	22	200	6.3
7	C	0.31	12.9	154	211	6.1	–	2475	24	200	1.3
Control	A	0.49	10.0	200	–	32.4	285	2315	23	100	6.6

- Most of the fly ashes studied exhibited a potential for use in the high volume fly ash structural concrete, although they had varying influence on the concrete properties both in fresh and hardened state.

- After one year of curing, compressive strengths with various fly ashes ranged from 42.7 to 83.4 MPa. In general, the increase in cement content from 155 to 225 kg/m^3 did not produce any significant gain of compressive strength in most of the high volume fly ash concretes.

- The creep strains up to 1 year loading period ranged from 156×10^{-6} to 352×10^{-6} in concretes with cement content of 155 kg/m^3 and from 211×10^{-6} to 501×10^{-6} in concretes with 225 kg/m^3 of cement. The HFCC mixes appear to produce lower creep deformation due to the presence of relatively larger fraction of unreacted fly ash particles in HFCC.

- Excessive dosages of naphthalene based superplasticizer resulted in increased setting times of certain concretes, particularly the mixes with high cementitious material content.

- The resistance of concretes to chloride ion penetration increased with age. The charge, measured in coulombs as per the AASHTO test T277-831 varied from 197 to 973 at 91 days.

Malhotra et al. (1990b) examined the durability of high volume Class F fly ash concretes and concluded that superplasticized and air entrained HFCC showed satisfactory durability against freeze–thaw attack, carbonation, chloride ion penetration, volume change, and alkali silica reaction. However, HFCC exhibited poor performance in deicing salt scaling tests. On the basis of their tests, they do not recommend the use of HFCC in

situations where concrete is to be exposed to repeated applications of deicing salts.

Standard tests were conducted to determine properties of
fresh concrete such as slump, unit weight, and air content, and
of hardened concrete such as compressive strength, flexural
strength, modulus of elasticity, flexural toughness, impact resistance, chloride ion penetration resistance, drying shrinkage, and
freeze–thaw resistance. The results showed that the HFCC mixes even with cement contents of 125 to 150 kg/m^3, the mixes
having total cementitious material at the levels of 350 to 450
kg/m^3 and fibre contents of 0 to 5 kg/m^3 had compressive
strength ranging from about 15 to 20 MPa at 28 days and from
about 40 MPa to 50 MPa at 365 days. The test mixes exhibited
excellent durability characteristics and had satisfactory workability and other characteristics for field applications. As a result
of these investigations, the polypropylene fibre reinforced
HFCC has been successfully used for shotcrete application on
rock outcrops, to cap coal waste dumps at different sites in Nova
Scotia, and to cover mine waste rock in Vancouver Island,
British Columbia, Canada.

Amongst the tested mixes, the concretes having cement content of 150 kg/m^3, a fly ash content of 250 to 300 kg/m^3, and
water to cementitious material ratio of 0.29 to 0.32 produced
both higher and consistent compressive strength results. Their
strengths were, on the average of the order of 7 MPa at 7 days,
21 MPa at 28 days and 42 MPa at 365 days (Malhotra et al.
1990a).

At the University of Calgary, a study was conducted to
develop optimum design mixes for high strength by incorporating Alberta fly ashes at the replacement levels of 40 to 60% by
weight of cement (Joshi et al. 1991, 1993). About 7 trial mixes
were tested using superplasticized and air entrained plain and
fly ash concretes with slump of 100 ± 20 mm and air content
of 6 ± 1%. In the mixes, cementitious material content was
varied from 380 to 466 kg/m^3, water to cementitious material

ratio from 0.28 to 0.36, coarse aggregate from 1012 to 1144 kg/m^3, and fine aggregate from 643 to 712 kg/m^3. The properties examined were workability, air entrainment, setting times, compressive strength, freeze–thaw resistance, alkali aggregate reaction, carbonation, and premature freezing performance. A summary of the mix design and some test data are presented in Table 4.4.

The results of study by Joshi et al. (1993, 1994) indicated that with fly ash replacement levels up to 50% by cement weight, concrete with 28 day strength ranging from 40 to 60 MPa and with adequate durability can be produced with a cost saving of 16% at 50% replacement level. Longer initial setting times, decreased early age strength, and inadequate resistance to scaling due to deicing salts, were some of the observed deficiencies of high volume fly ash concretes.

Recently investigations at CANMET (Malhotra et al. 1994) have led to the development of high volume fly ash concrete containing ASTM Class F fly ash, greater than 50% of the cementitious material, and polypropylene fibre up to 5 kg/m^3 for practical applications in shotcreting rock out crops. The workability of concretes made with water to cementitious material ratios of 0.29, 0.32 and 0.36 was maintained at 75 ± 25 mm slump by using proper dosages of superplasticizers. The mixes were air entrained with air content around 8% for adequate freeze–thaw durability.

7.3. HIGH STRENGTH FLY ASH CONCRETE

Ever since the adoption of Portland cement concrete as structural material in the 1900s, continual research and development studies in the fields of science and technology of cement and concrete, undertaken with clear objectives of improving its strength and durability, have resulted in remarkable achievements with regard to steadfast increases in its compressive strength. As of today, a few ready mix concrete producers have already batched and delivered concretes with 90 day strength of up to 131 MPa for the construction of some major high rise building in the United States. Continued and sustained research efforts are underway to develop concretes with ultimate strength at 90 days and beyond in the range of 172 to 207 MPa (Perenchio 1979). Researchers are aiming to produce special cement based material with strength as high as 731 MPa (ACI 1992). ACI Committee 363 Report (1992) and PCA Bulletin EB 114T (1994) on high strength concrete offer good ready references for the development history, engineering properties and various projects that have successfully used high strength concrete.

The present state of the art has established that under careful control and use of low water to cement ratios of 0.22–0.25, water reducing admixtures (superplasticizers) and mineral admixtures such as fly ash, blast furnace slag, silica fume, high compressive strength concrete can be consistently produced. Of course special handling placing and quality control techniques may be necessary to ensure the achievement of strength above 100 MPa at 90 days and beyond (Moreno 1990). Further research is required for optimization of mix proportioning, improved construction practices, special chemical admixtures, and appropriate testing methods for quality control, cost effective and quality cement based materials to develop concrete having strength of 150 MPa and more.

The incorporation of pozzolans including fly ash leads to progressive strength development with age because of the pozzolanic reactions. This property of fly ash concrete makes it a competitive candidate in the construction of structures requiring the use of long term high strength concrete with a low-heat generating capacity. High strength high volume fly ash concrete with 28 day strength greater than 50 MPa has been successfully used in foundations and lower storey columns of tall buildings, foundations of large bridges and power houses, offshore drilling platforms, and gravity base structures.

TABLE 7.5 Classifications of high-strength concretes
(Mindess et al. 1981)

Type	w/c ratios	28-day strength MPa (lb/in^2)	Remarks
Normal consistency	0.35–0.40	35–80 (5000–12000) high cement contents	50–100 mm slump
No-slump	0.30–0.45	35–50 (5000–7000)	less than 25 slump normal cement contents
Very low w/c ratios	0.20–0.35	100–170 (15000–25000)	use of admixtures
Compacted	0.05–0.30	70–240 (10000–35000)	compaction pressure 70 MPa (10000 lb/in.^{2}) and greater

According to the American Concrete Institute (ACI 1984), high strength concrete is described to have 28 day strength greater than 41 MPa, which is not yet hard and fast, as ACI also recognizes that the definition of high strength has to be flexible and can vary on a geographical basis. The tentative classification of high strength concrete as given in Table 7.5 and suggested by Mindess et al. (1981) appears to be somewhat relevant to the current state of the knowledge.

High strength concrete mixes incorporating fly ashes and with 28 day strength ranging from 55 to 70 MPa have been successfully used in the lower columns of high rise buildings in the Toronto area of Canada (Bickley and Payne 1979) and in the Chicago area of the United States (ACI SP-46-9, 1974).

Cook (1982) conducted an extensive study using a high calcium fly ash with CaO content of 30.3% at 25% replacement level to develop concrete mixes with 28 day strength in the range of 55 to 75 MPa. The results of his study are presented in Table 7.6. He reported that as cement factor was increased, sand content and the water to cementitious material ratio was

TABLE 7.6   Mix proportions and test results of high-strength test mixes (Cook 1982)

	Nominal cement content, lb/yd³				
	470	564	658	752	846
Materials					
Cement (Type I)	476	575	667	756	844
Fly ash (Class C)	119	143	167	189	211
No.57 limestone	1928	1934	1923	1905	1890
Sand	1238	1083	974	864	765
Water	250	264	272	285	301
WR-admixture (3 oz/cwt)	18 oz	22 oz	25 oz	28 oz	32 oz
Mix Properties					
Slump, inches	4	4	4.5	4	3.75
Unit weight, lb/ft³	148.5	148.1	148.3	148.1	148.2
W/C, lb/lb	0.42	0.37	0.33	0.30	0.29
Compressive Strength, psi					
7 days average	6419	6968	7693	7985	8595
28 days average	7914	9091	9774	10541	10177
56 days average	8840	9859	10964	11000	11309
90 days average	9179	9992	10876	10964	10876
180 days average	9823	11143	12115	12557	12361
Modulus of Elasticity, psi					
@28 days	–	–	5.71×10^6	5.98×10^6	–

1 lb/yd³ = 0.5933 kg/m³, 1 psi = 0.0069 MPa, 1 lb/ft³ = 16.02 kg/m³, 1 inch = 25.4 mm.

reduced. To maintain workability around 100 mm, water reducing admixture was used in all the mixes. The 180 day strength of the mixes ranged between 68 MPa and 86 MPa.

Swamy and Mahmud (1986) reported that concrete containing 50% low calcium Class F fly ash as cement replacement and a superplaticizing agent could develop 60 MPa compressive strength at 28 days and a fairly high strength of 20 to 30 MPa at 3 days. In a recent study, Joshi et al. (1993) demonstrated the development of concretes with 28 day strength ranging from 45 MPa to 60 MPa with 40 to 60% Alberta fly ash replacing cement and using superplasticizer and air entraining admixtures. It is now well established that fly ash through its pozzolanic reaction provides increased strength at late ages of 91 days and beyond that cannot be achieved through the use of just additional Portland cement in a mix.

With the judicious use of chemical admixtures and industrial by-products such as granulated blast furnace slag and condensed silica fume, it has been possible to develop high volume fly ash concrete with adequate early age strength. In this respect, the research investigation on fly ash, condensed silica fume and Portland cement concretes containing high range water reducer are of particular interest especially to the precast concrete industry for early strength development.

CHAPTER 8

Admixtures in Fly Ash Concrete

8.1. GENERAL

The basic ingredients of cement concrete are Portland cement, water and aggregates. Additives and /or admixtures are the substances that are used either as ingredients in blended cements or as additional compounds at the time of mixing to modify concrete properties. There are two broad categories in which admixtures for cement concrete can be classified: (1) chemical admixtures (2) mineral admixtures. Chemical admixtures such as accelerators, retarders, water reducers, superplasticizers or high range water reducers, and air entraining agents are soluble in water and react with cement. They are used in relatively small amounts by weight of cementitious materials. Mineral admixtures include natural pozzolans and industrial waste by-products such as fly ash, slag, silica fume, rice husk ash, and red mud. Unlike chemical admixtures they are used in relatively large amounts to substitute cement and/or fine aggregate in concrete. The properties, use, and influence of fly ash as mineral admixture have been described in previous chapters.

8.2. CHEMICAL ADMIXTURES

The most commonly used chemical admixtures in fly ash concrete are accelerators, water reducing agents, superplasticizers

173

(high range water reducers), and air entraining admixtures. Their role in fly ash concrete is described below.

8.2.1. Accelerators

Most fly ashes, except for some Class C fly ashes having self hardening characteristics, impart relatively slow setting and hardening characteristics to concrete. For practical use, particularly in strip form construction and for early removal of form work for both technical and economic reasons, it is some times necessary to accelerate the strength of fly ash concrete. Accelerators are generally inorganic salts of calcium, sodium or potassium. The most commonly used accelerator in fly ash concrete like normal concrete is calcium chloride. Chloride accelerators are available which meet the requirements of ASTM C494 Type E.

In a study on the effects of different types of accelerators and superplasticizer on concrete properties, Mailvaganam et al. (1983) found that both chloride and non chloride accelerators were effective in early age strength development of reference concretes. The results of the study are presented in Table 8.1. It is further reported by the authors that the slow rate of early strength development of fly ash concrete proportioned by simple replacement method substituting fly ash for cement on equal weight basis cannot be completely made up with use of accelerators.

In a study, Joshi et al. (1991) used a Darex set accelerator in a few selected trial fly ash concrete mixes. The admixture in the form of free flowing powder is of non chloride type, conforming to ASTM C494 Type E. They found that with the use of the accelerators at dosages of 0.5 to 1.0 kg per 100 kg cementitious material, no significant impact on the setting time or early strength development of concrete was noticeable. Larger quantities of set accelerators were not incorporated for economical reasons and the relatively insufficient gain in setting and hardening characteristics of high volume fly ash concrete tested.

TABLE 8.1 Properties of fly ash concretes incorporating chemical admixtures (Mailvaganam et al. 1983)

Mix no.	Admixtures	Compressive strength (MPa)				Shrinkage % x 10^{-3}	
		3 days	7 days	28 days	90 days	28 days	56 days
1	–	23.3	28.5	35.4	41.1	86	100
2	–	13.1	16.8	24.6	36.1	92	103
3	CL	29.7	35.3	40.5	46.9	89	106
4	CL	16.6	20.9	30.9	40.1	94	106
5	MFS	32.6	36.9	43.0	47.2	102	112
6	MFS	18.0	21.6	30.2	41.4	99	120
7	NCL	26.3	31.0	37.5	42.7	82	98
8	NCL	14.6	18.5	26.7	38.3	87	95

CL = chloride accelerator (1.5% active solids). MFS = superplasticizer (0.6% active solids).
NCL = non-chloride accelerator (0.6% active solids). All at a dosage of 3% by weight of cementitious material.

8.2.2. Water Reducing Agents

These are surface active agents derived from either lignosulphonic acids and their salts or hydroxylated carboxylic acids and their salts. Many of the water reducing admixtures are also set retarders. These agents are adsorbed on the cement particles, giving them a negative charge, which leads to repulsion between the particles and thus causing effective dispersion and deflocculation of cement particles. The particles have, therefore, a greater mobility and water freed from the restraining influence of the flocculated system becomes available to lubricate the mix so that the workability is increased. The addition of Class F fly ash in particular also improves the workability because of the spherical shape of its particles. Thus with water reducing admixture, water requirement to maintain workability in fly ash concrete is considerably reduced.

As reported by the U.S. Corps of Engineers, the use of water reducing admixtures in concrete containing fly ash produced relatively technically appropriate and economically cost effective concrete for structural applications. Lovewell and Hylond (1971) and Samarin and Ryan (1975) also reported satisfactory results when fly ash and water reducing admixtures were combined in making concrete mixes. The test data based on studies by Samarin and Ryan (1975) are presented in Table 8.2. The previous investigators, however, emphasize that trial mixes, containing the actual materials including fly ash to be used on the job, should be made in order to determine the type and quantity of admixtures for producing mixes with specified quality parameter.

Joshi et al. (1994) studied the effect of a water reducing agent, Daracem-100 which is an aqueous solution of chemical dispersants and some other chemicals, on the setting characteristics of a few trial fly ash concrete mixes. The use of this agent was found to delay the setting time of the fly ash concretes and was thus not pursued further. In view of the fact that water reducers may also act as set retarders, their use in fly ash

TABLE 8.2 Properties of concretes incorporating cement, fly ash, and water-reducing admixtures (Samarin et al. 1975)

| Series | Mix type | Cementitious content, wt % | | Water demand kg/m³ | Slump mm | Air % | Bleeding capacity kg/m² | Setting time (h:min) | | 28 days strength (MPa) | | Modulus of elasticity x 10⁴ MPa |
		Portland	Total					Initial	Final	Compressive	Tensile	
A	P	15.2	15.2	188	80	3.1	2.5	4:45	6:45	40.6	3.95	3.65
	F	14.1	16.7	182	90	2.3	2.2	4:40	6:15	44.6	4.38	3.61
	A	13.7	13.7	165	80	5.7	2.1	6:20	8:30	41.3	4.00	3.52
	F-A	13.1	15.7	170	80	5.2	2.0	6:30	9:05	39.8	4.10	3.41
B	P	19.1	19.1	196	80	2.4	0.9	4:10	5:35	47.5	4.49	3.68
	P	15.2	15.2	185	85	2.8	2.4	4:40	6:15	41.5	4.52	3.50
	P	12.2	12.2	193	80	3.8	4.9	5:30	7:30	27.5	3.56	2.98
	F-A	17.0	19.1	173	80	5.2	1.4	6:10	7:55	47.5	4.70	3.55
	F-A	13.1	15.7	169	80	5.3	2.5	6:30	8:25	39.7	3.78	3.30
	F-A	9.8	13.1	167	80	5.8	4.3	6:55	8:50	30.4	3.37	3.18

concretes which otherwise have slow rate of setting and hardening is not very popular in commercial concretes.

8.2.3. Superplasticizers

Superplasticizers also termed high range water reducers are highly effective as dispersive agents and assist in efficient hydration of cement particles at significantly low water cement ratios. Use of superplasticizers is helpful in developing high performance concrete with or without pozzolanic materials including fly ash. The present state of the knowledge has established that high strength high volume fly ash concrete can be developed by the combined use of superplasticizers, air entraining agents and highly reactive pozzolan such as condensed silica fume. With the use of superplasticizers and air entraining admixtures it has been possible to develop concrete containing fly ash up to 60% as cementitious materials with 28 day strength around 60 MPa. Such fly ash concretes possess adequate freeze–thaw durability and other types of characteristics. In fact the use of superplasticizers in HFCC has become routine to maintain the desired workability required for concrete placement at job site.

Superplasticizers can be classified into the following groups:

- Melamine based: sulphonated melamine formaldehyde condensate.
- Naphthalene based: sulphonated naphthalene formaldehyde condensate.
- Lignin based: modified lignosulphonates.
- Other surfactants.

Based on the results of the studies related to the effects of superplasticizers on the properties of concretes with and without fly ash, Lane and Best (1978) concluded as below:

Superplasticizers are compatible with fly ash in concrete and produce no detrimental effects. The benefits claimed for these admixtures in plain concrete, however, were not as apparent in fly ash mixtures, particularly with respect to compressive strength gains and duration of increased plasticity. Water reductions for equal slump did not exceed 15%, improving this characteristic only slightly over a standard water reducing agent. The low water reductions can be attributed to the lower water requirement for fly ash concrete as compared to plain concrete for equal consistency. Since there is less excess water initially available, the addition of water reducer is less effective.

Superplasticizers are equally effective in attaining a temporary increase in concrete consistency for both fly ash concrete and plain concrete. The highly plastic phase diminishes after 15 minutes and ceases after about 30 minutes with fly ash concrete.

It is worth pointing out that most of the studies on the effect of chemical admixtures on fly ash concrete were conducted using Class F ash. Since fly ashes vary significantly, all such observations should be verified for the particular fly ash used on a project.

It has been reported that the actual decrease in the mixing water depends on the cement content, type and quantity of admixture used, type of aggregates, presence of mineral admixtures or air entraining admixtures (Ericksen and Nepper-Christensen 1981, Brooks et al. 1982). Superplasticized concretes exhibit noticeable increase in slump, although these increases are of short duration. The effect of superplasticizer on slump extinguishes with 30 to 60 minutes after its addition to the concrete. The absence of effective slump increase due to superplasticizer presents some practical problems in the use of superplasticizer.

Swamy et al. (1983) developed flowing concrete, with slump of 260–280 mm, using 30% fly ash by weight of cementitious material and superplasticizer at 2.5% by weight of cementitious material. The air cured mixes had 28 day strength of 45.2 MPa compared to 48.0 MPa of 3 days water and 25 days air cured mixes. The 1 day strength of the fly ash mixes studied ranged between 10.4 MPa and 12 MPa comparable to that of plain concrete mix.

The test data in regard to the use of superplasticizers in developing high strength high volume fly ash concretes by Mukherjee et al. (1983) are given in Tables 8.3 and 8.4. The test mixes had cement content of 377 to 391 kg/m^3 and fly ash at 37% of cementitious material ratios of 0.28 and 0.35. The mixes were non air entrained. Three types of superplasticizers, M, N and L belonging to melamine, naphthalene and modified naphthalene groups, respectively, were added to maintain about 70 mm slump. The reference fly ash concrete with 37% fly ash was proportioned to have 28 day compressive strength of 40 MPa. The authors reported the following observations.

- Satisfactory high strength concrete can be obtained using large quantities of fly ash (low calcium) and various superplasticizer.
- The mechanical properties of the superplasticized fly ash concrete were superior to the reference concrete.
- The workability may impose a limit on its use for cast in place construction, due to a gluey texture at slumps between 65 mm and 74 mm.
- Superplasticizers N and L both increased the setting time markedly.

Malhotra (1981) reported the effects of naphthalene based superplasticizer on the properties of semi-light weight fly ash concrete. As presented in Table 8.5, the high strength of 36–38 MPa achieved at 1 day were probably due to the combined effects of fly ash and superplasticizer which allowed reduction in water cementitious ratio and without reducing workability.

TABLE 8.3 Mix proportions and properties of high-strength concrete (Mukherjee et al. 1983)

Mix	Batch	Mix proportions, kg/m³							Properties of fresh concrete		
		Cement	Fly ash	Fine aggregate	Coarse aggregate	Water	W/(C+F)	SP^a	Slump (mm)	Air^b %	Initial set h:min
1	1	377	223	420	1105	208	0.35		65	1.5	4:40
	2	378	224	420	1107	208	0.35		70	1.5	4:40
	3	377	223	420	1105	208	0.35		65	1.4	4:40
	4	377	223	419	1103	209	0.35		65	1.4	4:40
2	1	389	230	433	1139	171	0.28	M 1.97	75	1.5	4:25
	2	388	230	432	1137	171	0.28	1.97	70	1.7	4:25
	3	390	231	434	1141	172	0.28	1.97	65	1.6	4:25
	4	389	230	433	1139	171	0.28	1.97	65	1.6	4:25
3	1	390	231	434	1141	172	0.28	N 0.86	70	2.0	5:15
	2	391	231	435	1145	172	0.28	0.86	70	1.8	5:15
	3	389	230	433	1139	171	0.28	0.86	70	1.9	5:15
	4	390	231	434	1143	172	0.28	0.86	70	1.9	5:15
4	1	390	231	434	1143	172	0.28	L 2.70	75	1.6	7:40
	2	391	231	435	1145	172	0.28	2.70	75	1.4	7:40
	3	390	231	434	1143	172	0.28	2.70	75	1.5	7:40
	4	389	230	433	1139	171	0.28	2.70	75	1.6	7:4

a: % superplasticizer by weight of cement + fly ash. b: entrapped air only.

TABLE 8.4 Properties of hardened superplasticized concrete (Mukherjee et al. 1983)

Mix	Batch	Compressive strength of concrete (days), MPa						28 days flexural strength MPa	28 day modulus of elasticity $\times 10^4$ MPa	Shrinkage after 448 days dry storage	
		7	28	56	91	183	365			Shrinkage, %	Moisture loss, %
1	1		41.0		51.8			6.3	3.17	-0.0441	1.83
	2	28.8	37.9	51.9	51.5				3.43	-0.0453	1.84
	3	25.0		41.4				6.8			
	4					53.4	57.3				
2	1		53.3		65.2			8.3	3.48	-0.0413	1.19
	2	36.1	52.5	61.8	65.6				3.46	-0.0399	1.13
	3	36.1		57.7		70.0		8.0			
	4						69.8				
3	1		52.5		66.9			8.0	3.48	-0.0467	1.07
	2	36.8	53.8	61.7	67.1				3.46	-0.0427	1.08
	3	36.8		63.2		66.3		7.4			
	4						74.7				
4	1		51.0		62.7			7.8		-0.0455	1.19
	2	35.4	51.0	59.5	62.5				3.45	-0.0454	1.26
	3	34.1		59.0		66.4		8.1	3.45		
	4						62.5				

TABLE 8.5 Properties of semi-lightweight concrete incorporating fly ash and superplasticizer (Malhotra 1981)

Cement kg/m³	Fly ash kg/m³	Water kg/m³	W/(C+F)	Superplasticizer % on wt of cement	Strength, MPa 1 day	Strength, MPa 28 days
422		137	0.33	0.49	27.1	43.5
431		124	0.29	0.90	34.1	47.0
445		110	0.25	1.50	35.0	49.6
393	60	113	0.25	1.20	36.0	47.6
420	30	112	0.25	1.40	38.2	48.5

Sivasundaram et al. (1990) examined the properties of high volume fly ash concretes made with several fly ashes at 58% of the total cementitious materials, and using a naphthalene based superplasticizer and a synthetic resin type air entraining admixture. The test results are given in Tables 7.3 and 7.4. They reported that excessive dosages of the superplasticizer used to obtain the required workability around 200 mm resulted in delayed setting of concrete mixes with high cementitious material content. Similar observations were reported by Rodway and Fedirko (1992) in their tests on superplasticized high volume fly ash structural concrete for field use.

Joshi et al. (1991, 1993) reported the development of high strength high volume fly ash concrete mixes. WRDA-19, an aqueous solution of a modified naphthalene sulphonate containing no chloride, was used as superplasticizer in the test mixes to obtain workability of 100 ± 20 mm. The amount of superplasticizer used ranged between 800 and 1400 ml per 100 kg of cementitious material. Non fly ash reference concrete mixes required larger quantities of the superplasticizer to maintain constant workability at low water to cementitious material ratios of 0.28 to 0.36. An increase in fly ash content in the mix, the amount of superplasticizer needed to obtain the desired workability decreased as given in Table 4.4 (Akkad Salam 1992). Some increase in setting times with an increase in amount of superplasticizer in the mixes was also reported.

Uchikawa et al. (1982, 1983) examined the adsorption of three types of superplasticizers on low calcium fly ash and found that adsorption on fly ash was less than on cement in water suspensions. They attributed this behaviour to difference in the surface characteristics of fly ash and cement in water suspension. Nagataki (1983) reported that the adsorption of superplasticizer on fly ash appears to be reduced in the presence of carbon. The research investigations conducted so far do not indicate such abnormal and unfavourable interaction with superplasticizer or water reducing admixtures in practical concretes. On the contrary with fly ash addition, a relatively reduced amount of superplasticizer can produce concrete of the desired workability compared to plain concrete.

Ryan and Munn (1979) concluded from their study on the effects of fly ash and superplasticizer combination on slump loss that the superplasticizers are chemically affected by reaction with calcium hydroxide liberated during the hydration of Portland cement and thereby the loss of slump appears in the mix.

Under the present state of knowledge, it is found that the use of superplasticizer in fly ash concrete allows relatively high level of cement replacement by fly ash to develop concrete of specified consistency at lower water to cementitious material ratio and thereby higher strength and reduced permeability. The concretes so produced should be particularly suitable for applications in the situations where increased resistance to environmental and chemical attack is required. The interaction of fly ash with superplasticizer in the presence of other chemical and mineral admixtures incorporated in Portland cement concrete still need better understanding.

8.2.4. Air Entraining Agents

To improve freeze–thaw durability of fly ash concrete as is the case with plain concrete, minute air bubbles closely spaced at 0.20 to 0.25 mm are intentionally incorporated throughout the concrete mass by means of suitable agents, called air entraining agents (AEA) to achieve total air content of $6 \pm 1\%$. Numerous proprietary brands of air entraining agents are available commercially. The essential requirements of an AEA are that it rapidly produces a system of finely divided and stable foam, the individual bubbles of which resist coalescence. The produced foam should not produce deleterious effects on the cement hydration and other characteristics.

The AEA is either interground with cement in fixed proportions or used as admixture as an ingredient of concrete at the time of mixing. Significantly small amount of AEA, 0.005 to 0.05%, by weight of cement, are dissolved in water when used as an admixture for entraining the specified air content in the concrete.

Laboratory as well as field studies have revealed that the addition of some fly ashes causes an abnormal increase in the amount of AEA required to produce a given level of air entrainment in concrete. This problem has been of great concern to fly ash producers/vendors and concrete producers, because it is generally quoted as one of the governing factors for not allowing unrestricted and wide spread use of fly ash in concrete, particularly in the western Canada.

In the majority of previous studies, relating to the performance of concrete containing high volume of fly ash, a completely neutralized vinsol resin was used as an air entraining admixture in combination with naphthalene based superplasticizers (CANMET 1989, 1993). In the investigations in regard to the development of high strength durable concrete incorporating Alberta fly ashes at 40 to 60% replacement level, Joshi et al. (1993) used Daravair, an aqueous solution of completely neutralized vinsol resin, at 95–450 ml per 100 kg cementitious material along with superplasticizer to achieve an air content of around 6% in the concrete mixes. They reported that the mixes containing fly ash, in general, required 2 to 3 times more air entraining agent compared to the control plain concrete mixes. The increase of AEA varied in an abnormal and irregular pattern for different fly ash contents.

The erratic interaction of fly ash with air entraining agent has baffled the efforts of several researchers to explain the underlying mechanism. Based upon the test observations, it is generally accepted that fineness and carbon content expressed as percent loss on ignition (LOI) in fly ash play a predominant role in increasing the AEA demand of fly ash concrete. Air entraining agents are adsorbed on finely divided particulate other than cement.

Fly ash is normally finer than cement and the volume added is usually greater than the volume of cement replaced to produce fly ash concrete. Because of this surface area of the total constituents in concrete increases and thus a greater volume of AEA is needed to provide the same surface concentrations of the active air entraining ingredient. The second phenomenon possibly responsible for the increased AEA demand is related to the carbon content of fly ash. The carbon adsorbs a portion of the air entraining agent which makes it unavailable for creating the needed conditions for stable air bubbles. The amount of adsorption varies with the magnitude of carbon present and possibly with the form of such carbon. Therefore, variations in the carbon content or LOI result in a need to vary the amount of air entraining agent.

In Section 4.5, the effects of fly ash on air entrainment in fresh concrete was discussed based on the literature survey. Several investigators have reported that fly ash is not alone in causing increased AEA demand in fresh concrete. Other mineral admixtures or finely divided materials including some Portland cements behave in a similar way to require increased amounts of AEA for entraining specified air content.

The presence of organic constituents other than carbon may interact with the air entraining agent to reduce its effectiveness. In general, for structural applications high volume fly ash concretes also incorporate other mineral admixture like blast furnace slag or silica fume and superplasticizers. It is, therefore, pertinent to investigate compatibility between Portland cements, fly ashes, slag, silica fumes, superplasticizers and air entraining admixtures. The senior author is currently investigating the use of new AEA and ways to regulate and control the AEA demand in fly ash concrete as a function of some important characteristics of fly ash.

8.3. MINERAL ADMIXTURES

In addition to fly ash, ground granulated blast furnace slag and condensed silica fume are commonly used as supplementary cementitious materials. Because fly ash is a slow reacting pozzolan, the incorporation of fly ash as replacement for cement in concrete leads to the relatively slow rate of strength development in such concrete. Due to this reason, fly ash concrete can not be used for the structural applications where early strength is required. However, with the combined use of fly ash and other mineral admixtures such as slag and silica fume, the deficiencies that occur by only fly ash addition can be overcome.

Research in the past has established that ground granulated blast furnace slag develops strength in the presence of Portland cement more rapidly than most fly ashes do. Furthermore, it can be used to replace relatively larger amounts of cement, up to 70–80%, to produce satisfactory concretes. Triple blended cements containing fly ash, granulated slag and Portland cement have been studied by investigators around the world (Beal and Brantz 1992). Triple blended cements have been used on various projects in France (Fouilloux 1973) and in Australia (Samarin et al. 1983).

Condensed silica fume contains amorphous silica and is generated as a waste by-product from silicon metal and ferron silicon alloy industries. It is a highly reactive pozzolan and when added in small amounts, it allows development of higher early strength in concrete. Compared to fly ash, the addition of silica fume reduces the workability and thus requires increased water content to obtain the specified consistency. Use of superplasticizer is, therefore, needed to overcome this problem. The combined use of silica fume and fly ash in concrete is of more interest, as small amounts of silica fume addition in high volume fly ash concrete containing about 30% fly ash can improve early strength of concrete remarkably (Carette and Malhotra 1985, Mehta and Gjorv 1982).

Mehta and Gjorv (1982) studied the influence of silica fume addition on fly ash–Portland cement pastes. They measured pozzolanic activity from free lime content in paste of the mixes with and without condensed silica fume. The authors reported the increased pozzolanic activity of the mixes containing silica fume. Their study also showed that the presence of silica fume in Portland cement–fly ash mixes reduced the volume of large pores at all ages from 7 to 90 days.

An extensive investigation of concretes, containing 70% Portland cement and 30% low calcium fly ash with the addition of 5 and 10% condensed silica fume, was undertaken by Carette and Malhotra (1983). Superplasticizer was used for constant consistency without changing water content specially when silica fume was used. Based on the results, the authors reported that regardless of water to cementitious material ratio, strength at 1 day and 3 days was higher for concretes containing silica fume than for the control concrete containing 30% fly ash and 70% Portland cement, but was lower than equivalent concrete containing no additives. Pozzolanic activity with silica fume addition in fly ash concrete became more apparent between 3 days and 7 days whereas in fly ash concrete alone become noticeable after 28 days. Carette and Malhotra reported that at 20% silica fume addition, the slump reduced rapidly and the fresh concrete tended to be gluey and required larger quantity of superplasticizer to make it workable. An important finding from their study was that at ages beyond 7 days, the loss in compressive strength of concrete due to the incorporation of 30% fly ash generally can be completely compensated by addition of as little as 10% condensed silica fume.

Fly ash-condensed silica fume-Portland cement concretes have high potential for developing early strength concretes for use in structural applications, particularly in the precast industry and currently researchers are focusing attention to obtain more infirmation on their nature and long term performance. Of course, there is ample need for more research in investigating long term durability of concrete with different mineral combinations. The combined use of mineral admixtures to replace high volume of cement in concrete will lead to substantial cost and energy savings.

CHAPTER 9

Miscellaneous Opportunities for Fly Ash Use

9.1. GENERAL

Fly ash may be used as a raw material for the production of aerated concrete, also known as cellular concrete: a light weight material used to manufacture building blocks for residential, commercial and industrial structures. It is being used in autoclaved aerated, or cellular concrete (AAC or ACC) blocks produced by Celcon, Durox, Thermalite, Ytong and other European manufactures. In the United Kingdom, Celcon and Thermalite together utilized over 1 million tonnes of fly ash in this application in 1990. Typically, fly ash can account for 70 to 75% of the solid material that makes up ACC, and the Portland cement and lime make up the rest. A very small amount of aluminum powder is introduced in the slurry mix to create the hydrogen bubbles that impart the light weight cellular quality to this product. ACC is steam cured in a high temperature, high pressure autoclave, which gives the material outstanding dimensional stability and other desirable engineering characteristics as building material (PCA 1991).

Several attempts, including the recent one by Virginia based North American Cellular Concrete (NACC), have been made to introduce ACC or AAC manufacturing plants in the United

States, especially for the manufacture of blocks for use in masonry construction (PCA 1992), Electric Power Research Institute (EPRI) has identified autoclave cellular concrete as a potentially larger user of the 75 million tons of fly ash produced annually by its sponsoring coal burning utility companies. A mobile demonstration plant to produce fly ash based cellular concrete blocks from the fly ash has been designed by NACC. Demonstrations at various utility plants will include production and usage technology of fly ash based cellular concrete blocks to architects, builders, masons, developers, inspectors and the general public. NACC in collaboration with some European ACC companies are making sustained efforts aimed at providing an expanding U.S. market for this unique building material.

9.2. FLY ASH BRICKS

Coal ash has been used as a component in building bricks in Eastern Europe as early as in the 17th century. The major properties of fly ash exploited in the clay brick industry are:

- similar composition as that of clay
- fuel value due to the presence of unburnt carbon
- reduced weight of the resultant product
- reduced shrinkage due to its inert nature
- chemical compatibility with natural clays

Fly ash can be used to replace up to 40% of the raw material, clay, in building blocks and tiles (Bizen et al. 1991). It can also be used as a filler for clays which are too plastic to reduce the drying shrinkage of the products.

In the semi dry process used for manufacturing bricks, the mixed raw materials including fly ash and clay are compacted under high pressure, typically 10 to 40 MPa, to produce the green products, which are subsequently dried and fired. Clay is added to the fly ash as a binder to produce a green product with adequate strength for handling purposes. The unburnt carbon in the fly ash provides part of the process heat during the manufacture of bricks. Fly ash has also been used as a partial or total replacement of quartz sand in the production of sand–lime building bricks by using autoclave process.

In the early 1970s, the International Brick and Tile Company of Edmonton, Canada, set up a factory to produce brick, block and tile from fly ash using U.S. technology. Because of lack of technical and environmental support as well as economic factors the plant did not reach the commercial stage and has since been abandoned. In a similar way, several laboratory and pilot programs were initiated in the 1950s and 1960s in Europe and the United States to exploit the advantage of using fly ash in brick production. However the commercial production was not technically sound and cost effective. Major problems associated with fly ash use in the brick industry include dusting during handling, the lack of natural plasticity of ash, and the presence of soluble salts. Recent research in Britain and North America indicates that with the use of carefully selected and controlled ash, bricks with improved engineering properties can be produced at an economical cost (Clarke 1993).

9.3. MINERAL WOOL

Mineral wool is a fibrous product made from melted rock, glass and slag for use as an insulating material in the walls of building in cold region for pipes and also sound proofing. Fly ash, bottom ash, boiler slag, and so called modified coal ash obtained from circulating fluidized beds, are all potential raw materials for mineral wool manufacture. Of primary interest in the manufacture of mineral wool is the attainment of a suitable viscosity in the melt at the lowest possible temperature.

Dockler and Manz (1993), at Energy and Environmental Research Centre, University of North Dakota, reported that coal ash comprising 30 to 40% fly ash produced by wet bottom

TABLE 9.1   Chemical composition
range of mineral wool

Constituent	Weight %
SiO_2	35–65
Al_2O_3	0–33
CaO	5–50
MgO	0–32
Fe_2O_3	6–26

boilers in many power plants throughout the world can be used as a raw material for good quality mineral wool. At the present time, almost all the mineral wool produced in the United States utilizes slag obtained from steel mills as a primary raw material. Additives such as limestone or dolomite are typically used to adjust the chemical composition of the mineral wool products. The mineral wool is produced by placing the raw material and the additives into a Cupola furnace using coke as fuel. The processing generally includes (1) melting the feed charge (2) fiberizing the molten matter by using spinning or blowing methods (3) manufacturing of various mineral wool products, such as loose wool or insulation belts. A typical mineral wool has the range of chemical composition as shown in Table 9.1.

Pulverized coal is fed at high velocity and combusted in wet boilers of power plants to generate steam for electric power production. Two types of combustion residue, fly ash and slag or bottom ash, are generated as a result of coal burning in these furnaces of the power plants. About 60 to 70% is collected as slag and the rest as fly ash in the wet bottom furnaces. The molten slag falls to the bottom of the boiler and collects in a pool. The pool of molten slag, at temperatures of about 1425°C (2600°F) is drawn off through an opening in the bottom of the boiler, quenched in water and then slurried out of the power plant for disposal. The availability of molten slag from the wet bottom furnace affords highly desirable and economical raw material along with some heat energy necessary for fiberization for the production of mineral wool. The stream of molten slag from the wet bottom furnace is deflected to a fiberization device for necessary processing to produce mineral wool.

Preliminary fiberization tests using slag of the Coyote power plant, in the North Dakota state of the United States has demonstrated promising opportunity of producing mineral wool from coal combustion residue of wet bottom furnaces. For maintaining the particular chemical composition of the product, it may be necessary to install a device to add some type of fluxing agent to the melt such as lime, limestone, or dolomite. The need for flux may possibly add to the cost of the ash-based processes for mineral wool production. However, by selecting and controlling the chemical make up of the feed coals, the desired acid/base ratio ($SiO_2 + Al_2O_3/CaO + MgO$) in the range of 0.8 to 1.2 of the resultant slag can be maintained. In addition, the energy already in the molten slag would result in energy and cost savings which would partly make up the added costs for adding fluxing agents.

9.4. GYPSUM WALL BOARDS

A study was conducted to examine the technological and economic feasibility of producing building wall boards by using fly ash as partial replacement for gypsum (Joshi and Thomas 1990). The test data suggest that up to 30% gypsum may be replaced by fly ash without any significant change in the properties and service performance of wall boards. The fly ash use also makes the boards lighter. The improved thermal stability and insulating properties due to the addition of fly ash in the wall board may aid in reducing energy requirements. Thus, the partial replacement of gypsum by fly ash in wall boards is likely to affect considerable saving in energy consumption and in the conservation of natural resources. In addition, environmental benefits result from utilizing the amount of coal ash, a waste product, which must be disposed in ponds or land fills. The industrial production of fly ash–gypsum wall boards for commercial use is still in development stage.

Yet another type of coal combustion by-product which has considerable potential of direct use in wall boards is the by-product gypsum obtained from flue gas desulphurization process in coal burning power stations employing wet exhaust gas scrubbers. Future research and development studies are needed to establish the technical viability and economic feasibility of commercial use of by-product gypsum, oxidized flue gas desulphurization (FGD) waste, in wall board production.

9.5. LEACHATE MIGRATION CONTROL

Polymer, polyvinyl alcohol (PVA) and/or acrylic emulsions, modified fly ash and lime/cement mixes have been studied in laboratory using bench scale models to construct impermeable, flexible, and adequately durable 30–45 cm thick curtains in the form of intersecting cylinders (Joshi et al. 1993). These continuous curtains in the form of rings may be effectively and successfully used in arresting and even eliminating migration of leachates or pollutants from the old or existing waste dumps, sludge ponds, and/or from new oil or commercial spills. High volumes of fly ash can thus be utilized and effectively disposed for environmental protection purposes with added economic benefits.

9.6. OIL-WELL CEMENTS

In the late 1920s, oil and gas industry introduced the process of oil well cementing with clear objectives of:

- protecting oil producing zones from salt water flow,
- protecting well casings from blow out and collapse under pressure,
- protecting well casings from corrosion,
- reducing the dangers of contamination of ground water by oil, gas or salt water,
- bonding and supporting the casing.

For oil recovery, a bore hole is first drilled. Then within it a mild steal casing is lowered. Cement slurry is then poured into the casing. After the slurry is poured, a rubber plug is placed and pressure applied on top of the plug by means of a water column. The idea is to pump the cement–water grout within the casing through the bottom of the casing to the annulus around the casing, in the open hole or into fractured formations. Once the cement–slurry is pumped, it is allowed to set for sometime and then only those strata that have oil are perforated to recover the oil. The cement effectively seals off all the other strata of the formation so that gases and water do not contaminate the oil bearing strata.

Oil well cements are, generally, manufactured in accordance with the specifications of American Petroleum Institute (API) Standards 10A "Specifications for Oil Well Cements and Cement Additives." They are used at temperatures ranging from well below freezing at the surface to above 370°C or higher in fire–flood wells, and to depths greater than 9,000 m (30,000 ft). Pressures as high as 130 MPa may exist in deep hot wells. Specifications do not cover all the properties of cement over such broad ranges of temperature, depth and pressure. More than 40 additives, organic and inorganic and including clays and related materials are used to meet the wide range of conditions and requirements of oil well cements.

Fly ash was identified as a suitable admixture for use with Portland Cement to produce suitable cementing materials for

TABLE 9.2   API specifications for fly ash

Physical properties		Chemical analysis (%)	
Specific gravity	2.46	SiO_2	43.2
Weight equivalent in absolute		$Fe_2O_3 + Al_2O_3$	42.93
Volume to 1 sack (94 lb) cement, lb	74	CaO	5.92
Amount retained on #200 mesh sieve (75 μm), %	5.27	MgO	1.03
Amount retained on #325 mesh sieve (45 μm),%	11.34	SO_3	1.07
		CO_2	0.03
		Loss on ignition	2.98
		Undetermined	2.21

oil/gas wells as early as in 1949. API specifications for fly ash for use in oil well cements are given in the Table 9.2.

Fly ash–Portland cement mixtures are generally recommended for primary cementing at temperatures above 60°C (140°F). A compressive strength of 3.5 MPa is considered more than adequate to support casing and forms the basis of many regulatory specifications for cementing wells. Strength development in well environment of fly ash–cement is influenced by time, pressure and temperature.

As pozzolanic reactions are favourably influenced by pressure and temperature, strength increases and ultimately approaches or exceeds that of Portland cement.

The cement slurry is heavy, as a result the pressure required to pump it through the annulus is very high. Also due to the higher specific gravity of Portland cement, in the range of 3.14–3.15, the cement slurry tends to migrate in between the strata and soil formation rather than rising up in the annulus. This tends to create problems while making perforations in the oil bearing strata. These problems can be avoided by making the cement slurry light by adding fly ash as partial replacement of cement. Fly ash, due to its pozzolanic properties and its lower specific gravity of 2.0–2.5, is an ideal substance for using with cement for this operations.

Cements used in wells must exhibit a long life under corrosive conditions in the bore hole. The reaction of fly ash with calcium hydroxide liberated during hydration of Portland cement makes fly ash–Portland cement mixture less subject to the leaching action of corrosive waters and maintains low permeability over a long period. The leaching of calcium hydroxide, a vulnerable compound, to salt bearing waters is reduced by pozzolanic reactions. The permeability of hardened fly ash–cement slurry cured under down hole conditions is often less than that of the formation across which the cement is being placed.

Fly ash contains hollow spherical particles termed cenospheres or floaters having specific gravity less than 1.0. Ultra low density cement grouts of 1000 to 1300 kg/m^3 suitable for oil/gas well cementing have been formulated using such high strength cenospheres. The relative density of slurries can be reduced further to about 630 kg/m^3 by proper mix proportioning. Thus both fly ash and Cenospheres derived from fly ash are widely accepted as suitable additives for producing cementing materials in the oil/gas industry. A substantial amount of fly ash, produced in the Prairie provinces of Alberta and Saskatchewan in Canada, is being used for oil and gas well cementing. This use of fly ash by the oil industry is particularly advantageous, as much of the material is supplied during the winter months when other construction requirements are minimal.

The merits of fly ash incorporation for partial substitution of Portland cement in oil well cements are ease of retardation, light weight, reduced permeability, strength stability at high temperatures and pressures, and economy. This use provides yet another cost effective and environmentally friendly utilization of coal combustion by-products.

9.7. LIGHT WEIGHT AGGREGATES

Because of its unique properties, fly ash has been used for many years in Europe, particularly in the United Kingdom, to produce light weight aggregate. Sintered light weight fly ash aggregate under the trade name Lytag, manufactured by the Lytag Company of Britain, is commercially marketed for structural applications in building industry in the production of concrete masonry units and other building components.

The production of aggregate from fly ash involves (1) pelletizing of the ash into the small pellets of specified size (2) sintering the pelletized ash at 1000 to 1200°C in shaft kiln, rotary kiln or sinter strand. In the present Lytag process, a travelling grate sinter machine is used for firing. The sintered nodules provide a very good aggregate of specified shape and glassy surface texture. The bulk density of aggregate produced is about 960 kg/m^3.

The various attributes which allow fly ash utilization for manufacture of light weight aggregates are:

- The small amount of unburnt carbon in Class F fly ash may contribute substantially to the fuel demand of the sintering process.
- Reactivity of fly ash with water may facilitate the formation of good quality green pellets.
- The chemical composition is similar to that of clays.
- The glassy phases present in fly ash may reduce energy demand of sintering process.

In the early 1970s, Enercon Ltd. of Toronto made some efforts to co-produce light weight aggregate and quality pozzolans with concurrent separation of other high value elements from the ash by installing an ash beneficiation plant at the Lake View generating station. Although no significant commercial production occurred at the plant, since it was dismantled, it can be taken as a pioneering effort in Canada to produce light weight aggregate from fly ash.

Currently interest appears to be growing rapidly both in the United States and Canada towards total resource recovery from fly ash with emphasis on increased utilization of coal combustion by-products by the Construction Industry.

In the United States, the Poz-O-Tec process is currently in operation at over thirty locations for the production of light weight aggregate using wastes from wet scrubbing by lime, limestone, or dual alkali scrubbers (Smith 1993a). A commercial product, Poz-O-Lite, is a combination of bituminous coal fly ash, flue gas desulphurization (FGD) sludge, and lime.

Under the environmental regulations of clean air, the majority of coal burning power stations in the United States have adapted the measures to scrub the exhaust gases resulting from coal burning so that the primary contaminants, sulphur oxide (SO_x) are reduced to the permissible regulatory level. Oxidized limestone scrubbing produces gypsum $(CaSO_42H_2O)$ of uncertain quality which can be effectively fixated/stabilized with fly ash and lime. Nearly 40 million tons of fixated fly ash FGD waste is being produced annually in the United States. In order to exploit the chemical potential of pozzolanic cementation of the fixated waste, commercial production of light weight aggregate with no heat or low heat has been successfully demonstrated in 1992 (Smith 1993a).

The majority of light weight aggregate including the new Poz-O-Lite aggregate derived from fly ash and FGD waste is utilized in the production of concrete masonry units with the resulting benefits of lower transportation costs, lower laying costs, and superior insulating, acoustical and fire resistance characteristics. The aggregates produced are required to comply with the specification of ASTM C-331 "Standards Specifications for Light Weight Aggregates for Concrete Masonry Units." Recent applications of the new Poz-O-Lite aggregates have established the technical and environmental feasibility.

9.8. RESOURCE RECOVERY FROM COAL ASH AND OTHER USES

Coal burning power stations on global basis generate yearly millions of tonnes of wastes, including fly ash, bottom ash, boiler slag, and flue gas desulphurization sludge. At present, only a small proportion of the coal ash produced is used commercially, the rest is disposed of in ponds and landfills with attendant environmental problems and added cost to the utility industry and thus the consumers. As of this writing, coal fly ash, fine fractions of coal combustion residue, is used in structural fills, backfills, road bases, soil stabilization and asphalt concrete, snow and ice control, blasting grit/roofing granules, grouting, mining applications, land reclamation, and in preparing bricks, blocks, tiles, Portland cement, pozzolan etc. In addition to these uses, fly ash can be used for treatment of waste waters and for conditioning, stabilizing/solidifying waste sludges.

Fly ash also has been recognized as a potential source of multiple raw materials for other process industries. A variety of processes have been developed for the recovery of resource materials from coal residues, including aluminum, Vanadium, magnetic iron, titanium, activated carbon, cenospheres, mineral wool, and fine particulate fillers for ceramic, glass and plastic industry. Most recovery processes, although technically feasible, are currently uneconomic compared with other available processing resources and methods.

In the mid-1970s, energy costs jumped drastically and concern over declining resource supplies resulted in growing interest in the development of integrated materials recovery, and total resource recovery from coal ash. A critical appraisal of resource potential of fly ash reveals that components such as magnetic iron concentrate (magnetite), aluminum, cenospheres, or hollow light weight spheres, can be extracted at reasonable cost together with silica and aluminosilicates which can be used as fillers or raw materials in various industries.

Fly ash has a number of unique characteristics which make it marketable as a filler or extender in asphalt, plastics, glass as well as rubber industries. Some of these are:

- Fly ash can be brought to the market without the high energy costs associated for mining, processing and pulverization of traditional filler materials.

- Fly ash particles are uniquely or predominantly spherical, they impart improved packing and rheological properties to the composite, mortars and plastics.

- Fly ash particles have high compressive strength and good thermal stability, which make them suitable for high temperature applications including refractory and specifically ceramic materials.

There are basically two advantages which relate to the selection of fly ash for extracting magnetic iron concentrations, cenospheres, mineral wool and filler products. The first is that recovery process for these resource materials are relatively simple and well developed and require little processing beyond physical separation. The second is that existing markets are well established for competing products or currently import fly ash-derived materials. Potential applications of magnetic iron concentrate derived from coal ash include synthetic magnetite for coal washing, iron ore supplement, and iron salts for sewage coagulation. The largest market for magnetic concentrate is, however, for coal beneficiation by heavy medium separation.

Fly ash derived cenospheres, hollow sphere with specific gravity <1.0 can be used as substitutes for manufactured hollow glass spheres in the following applications:

- Ultra low density cement grouts for oil and gas will cementing work with relative densities ranging from 630 to 1000 kg/m^3.
- Light weight refractory materials. Cenospheres can be used as refractory aggregates along with refractory cements or sintered without any binders.
- Mineral filler. Because of their high strength, low weight, chemical inertness and particle size distribution, fly ash-derived cenospheres provide low cost substitutes for manufactured glass microspheres and other low density fillers. Their use in conjunction with binders such as organic resins or Portland cement renders desirable engineering properties on the finished product.
- Special aluminum alloys called "Ashalloys" containing cenospheres. They are much more resistant to heat and are many times stronger than the present day alloys.

Other products such as aluminum, activated carbon, mineral wool, ferro-silicon and trace metals are extractable provided economic factors favour development of suitable recovery processes.

CHAPTER 10

Fly Ash Usage in Waste Management

10.1. WASTE WATER MANAGEMENT

Most fly ashes possess certain unique characteristics which make it suitable for use in treating polluted water. The important properties are:

- The adsorptive properties of fly ash.

- The water extracts of lime and gypsum in some fly ashes effectively precipitate inorganic phosphorous and neutralize acid conditions in waste waters.

- The fine particulate nature of fly ash and the resultant high specific surface allow its use as a conditioner of waste water sludges for dewatering of vacuum filtration (Tenney and Echelberger 1970).

In recent years there have been many reports of the use of fly ash to remove trace amounts of heavy metals such as lead, nickel, mercury, chromium, cadmium, zinc and copper from aqueous waste streams (Panday et al. 1985, Sen et al. 1987, Weng 1990, Viraraghavan et al. 1991, Nonavinakere and Reed 1993).

205

Heavy metals present in industrial waste streams are toxic to humans and can have a long term adverse environmental impact. The potential sources of heavy metals include effluents from electroplating, metal processing, storage battery manufacturing, dyes/pigment manufacturing, tanning industries, leachate from landfills, and land applications of heavy metal-laden sludge. The acid–base characteristics of fly ash make their use in the treatment of metal bearing waste streams and the stabilization of metal sludges especially effective. Fly ash has been used as a coagulant to aid the settling of nickel flocs in precipitation treatment of spent electrolysis nickel (EN) plating baths which contain more than 6,000 mg/l of nickel hydrophosphate, phosphite, and/or many of the organic chelating agents such as citric, tartaric, lactic, succinic, and glycolic acids. Coal ash, thus can be used in the following ways for waste water treatment:

- as a flocculant or flocculant aid;
- as a neutralizing agent;
- as absorbent and adsorbent because of its high specific surface area; the presence of unburnt carbon in fly ash may provide added advantages in this respect;
- as a pozzolanic material, in the presence of lime, it acts as stabilizing/solidifying agent for waste sludges.

10.1.1. Adsorption

Removal of heavy metals by adsorption at solid–solution interface is a feasible means of treating metal bearing waste streams. Various materials such as activated carbon, metal oxides, and ion exchange resins have been used to remove trace metals from industrial and drinking water streams. Keeping in view cost and availability of adsorbents, fly ash because of its large surface area and carbon content may provide an economically viable and technically efficient alternative for adsorption of organic material (Johnson et al. 1966). Tenney and Echelberger (1970) report that fly ash can also be used for the effective removal of colour, odour, bacteria and algae from waste water streams. The fly ashes which can not be used in cement concrete because of excess carbon may be effectively used for cleaning the waste water. The adsorptive capacity is a function of the amount of unburnt carbon present in fly ash. The adsorptive capacity of coal ash can be obtained from the Freundlich's adsorption isotherms using the equation

$$\frac{X}{M} = KC^{1/n},$$

where X is the weight of organic material removed, M is the weight of adsorbent material, C is the residual concentration of dissolved organic material, and K and n are equation constants.

10.1.2. Water Extracts

Elements that have been reported to be associated with fly ash in trace amounts include silver, arsenic, boron, barium, chloride, cadmium, cobalt, chromium, copper, mercury, fluoride, manganese, molybdenum, nickel, lead, antimony, selenium, strontium, titanium, vanadium and zinc. The major components of fly ash comprise mostly silica, alumina, crystalline quartz and oxides of iron, calcium, magnesium, titanium and sulphur. The type and amount of metal oxides that form on the particle surface is a function of the type of coal and combustion used. Also whether a fly ash is alkaline, neutral or acidic in nature is determined by the type of coal and the air pollution control method used.

As indicated earlier it is worth mentioning here that the chemical composition of fly ash and many other materials is reported in terms of oxides. It does not necessarily mean that various elements are present as oxides.

Certain components of fly ash such as anhydride ($CaSO_4$) and lime (CaO) are water soluble. Brink and Halstead (1956) report approximately 1.8% release of $CaSO_4$ and 1% of CaO along with some sodium and potassium hydroxides during mixing of fly ash with water. The hydroxide ions are released from acid waters. The solution of hydroxides in water gives an alkaline environment or high pH. Lime and gypsum will dissolve to increase the calcium ion concentration, whereas sulphate ions in gypsum increase the sulphate concentration of the water. The calcium ions present remove phosphorous from waste water which is responsible for blooming or eutrophication. Because of this characteristic, fly ash is used for cleaning up lakes in which runoff from agricultural lands is discharged.

The oxidation of ferrous ions in fly ash in the presence of oxygen and hydroxyl ions tends to lower the pH. Also, the addition of calcium from fly ash into a water body tends to increase the hardness of water, though it is generally within acceptable limits. Thompson (1963) reports that the addition of fly ash up to 10 g/l to water bodies does not seem to cause any adverse effect on aquatic life.

10.2. ACID MINE DRAINAGE (AMD) NEUTRALIZATION

Alkaline fly ashes provide an effective means for treating mine drainage of acidic substances by the process of chemical neutralization of the acids through its hydroxides, and alkaline components. A recent study by Daniels et al. (1993) has indicated that acid forming coal refuse and alkaline fly ash can be "co-disposed" of in a way to balance refuse acidity by their bulk blending without causing other secondary water quality effects and with significant environmental and economic benefits. Earlier, Tenney and Echelberger (1970) demonstrated significant reduction of acidity from mine drainage by incorporating fly ash in the mine discharge. Acid mine drainage (AMD) with higher iron content, corresponding to higher acidities, has a higher buffering capacity. This means higher dosage of fly ash need to be applied for complete neutralization.

A good AMD control scheme has a relatively short and rapid mixing of the acidic discharge with fly ash and then separation of solids and liquids for effective neutralization. Compared to more conventional lime, top soil and rock phosphate treatments, alkaline fly ash amendment using optimum amounts of fly ash may be more effective in neutralizing AMD production in the acid producing natural soils or coal refuse. Considerable research and development to establish the effectiveness of AMD control of fly ash addition is, however, necessary for practical applications.

10.3. CONDITIONING, SOIL STABILIZATION AND SOLIDIFICATION OF WASTE SLUDGES

In recent years, many studies have been conducted to evaluate the use of coal ash, particularly fly ash, in the conditioning, fixation, stabilizing and solidification of sludges produced by biological treatment plants, municipal and industrial waste waters, oil/gas well drilling and scrubber or FGD. The waste sludges are in semi liquid state and some may be toxic. The present environmental regulations not only in the United States and Canada, but also in many other industrially developed countries, require that the waste sludges be disposed of in an environmentally acceptable manner. Class F and Class C fly ashes can be used for this purpose. In some cases Class F fly ashes are preferable, whereas in other cases high calcium Class C fly ashes are useful.

10.3.1. Waste Water Sludge Conditioning

One method of sludge handling and disposal is by dewatering the sludge by vacuum filtration followed by incineration of the dewatered cake. The dewatering is done to remove sufficient water so that the heat required for incineration and removal of the rest of the water from the sludge is minimized. Biological sludges, because of their high compressibility require pretreatment (Katsiris and Kouzeli 1987, Elini and Amodie 1993).

Sludge preconditioning means additions of finely divided materials and/or chemicals to the sludge mass for the purpose of forming a filter cake with sufficient strength and rigidity to withstand the pressure generated by vacuum. The addition of fine particulate from fly ash gives sufficient strength and rigidity to the filter cake. For optimum conditioning, sufficient quantities of fly ash must be added to support the sludge particles present in the filter cake. Since fly ash has a high absorptive capacity for water, an increase in fly ash content will also decrease the amount of water needed to be removed or filtered.

In a laboratory study on conditioning of municipal and industrial waste water sludges, Elini and Amodie (1993) observed that the most promising dosages for the sludge conditioning were·

- 1:1 of sludge/fly ash with 18% lime,
- 1:0.5 of sludge/fly ash with 18% lime.

The use of fly ash as a true "chemical reagent" in the dehydration process of municipal and industrial sludges indicates that it is technically feasible and economically viable as compared to the utilization of conventional reagents, such as iron chloride and lime or polyelectrolyte and lime.

The disadvantage of using fly ash is that, increasing the fly ash content increases the volume of filter cake to be incinerated. Nonetheless, this cake has better characteristics then those which do not contain fly ash. It has higher porosity, lower residual moisture content as well as high fuel value due to carbon content of the fly ash. The usage of fly ash as a conditioner also decreases the amount of organic and phosphorous pollutants in the filtrate water (Tenney and Echelberger 1970).

10.3.2. Scrubber Sludge Solidification

In the late 1960s, the enactment of Clean Air Legislation and subsequently the Clean Air Act Amendments of 1990 in the United States, the majority of the coal burning power stations have adopted wet scrubbing of exhaust gases for reducing sulphur oxide emissions to the permissible regulatory levels. Lime or limestone slurries are used for scrubbing and the waste products obtained are either a fine particulate suspension of calcium sulphite semi hydrate ($CaSO_41/2H_2O$) under unoxidized conditions and/or much larger by-product in the form of calcium sulphate dihydrate or gypsum ($CaSO_42H_2O$) as a result of forced oxidation during or after scrubbing. Gypsum from the oxidized by-product exhibits many advantages over unoxidized waste in that the dewatering of the sludges leads to a mass which is much easier to handle besides having some commercial use. Also, the oxidized by-product can be physically stacked or pounded without the extreme quicksand like characteristics exhibited by unoxidized by-products in their natural state (Smith 1993b).

A scrubber sludge, from a typical power plant containing 5 to 15% solids in suspension, undergoes a primary dewatering step via thickeners or hydrocyclones which raises its solid content to 30–50%. A subsequent secondary dewatering stage, most often by rotary drum vacuum filters or centrifuges, increases the solids content to 60–80%. The resulting filter cake, or centrifuge cake, is blended with the fly ash produced at the power station and a small percentage of lime. As a result of pozzolanic and sulpho-pozzolanic chemical reactions over a period of time, a monolithic mass with compressive strength of 6 to 12 MPa (850 to 1600 psi) and permeability coefficient of 1×10^6 to 5×10^9 cm/sec is obtained. This method of fixating/stabilizing FGD or scrubber sludge by using fly ash and lime is called "Poz-O-Tec" and is in operation at over thirty locations in the United States (Smith 1993a). It is estimated that in the United States nearly 50 million toms of the FGD stabilized sludge will be produced annually utilizing slightly above 15 million tons for fixation/stabilization of scrubber sludge (Smith 1993b). A substantial amount of the Poz-O-Tec material, fly ash lime-stabilized FGD sludge, is utilized in highway construction and is now being employed in commercial production of synthetic light weight aggregate with a trade name "Poz-O-Lite."

10.3.3. Oil/Gas Well Sludge Solidification/Detoxification

Joshi et al. (1995) recently examined the feasibility of stabilizing/solidifying oil/gas well sludges using fly ash and lime/Portland cement mixtures. Oil well sludge samples comprising drilling mud and formation cuttings as well as various hydrocarbons were collected from oil fields of Alberta, Canada. They were mixed in various proportions and combinations with the solidifying materials to determine an optimum formulation mixture. Compressive strength, permeability and microtoxicity tests were conducted to evaluate the effectiveness of solidifying agents.

The test results indicated that oil/gas well sludges with solid contents of more than 30% and having moderate toxicity can be effectively solidified/stabilized using a mixture of cement and fly ash in equal proportion at the rate of 15% by weight of solids in the sludge. Addition of 1% sodium silicate by weight of the cement accelerates the strength development. Solidification also reduces the toxicity of sludges significantly.

CHAPTER 11

Special Problems Including Use Constraints

11.1. ENVIRONMENTAL IMPACT OF EMISSIONS FROM PULVERIZED COAL BURNING POWER STATIONS

Since the oil embargo and adoption of Resource Conservation Recovery Act (RCRA) enacted in 1976, emphasis on worldwide basis including North America has been to burn more coal and less petroleum fuel for power generation. The basic attraction of coal as primary fuel is attributed to its low cost and widespread availability in abundance throughout the world. For 1989, The International Energy Annual indicates that the total coal production worldwide, was 4636 million tonnes (Michalski 1991). The largest producers in descending orders were China, the United States, the former USSR, the former East Germany, and Poland. Approximately 562 million tonnes of coal ash were produced in 1989 on a worldwide basis by coal-fired power plants.

In North America the biggest hinderance to expansion of coal-fired power generation is the cost of environmental protection due to Clean Air Legislation introduced in the 1960s and for complying with Clean Air Act Amendments of 1990. In the United States, for example, as much as 40% of the capital cost and 35% of the operating costs of new coal burning power plants can be accounted for by their pollution control systems. A brief description of air pollutants produced by coal-fired power plants is given below.

During the pulverized coal combustion in suspension-fired furnaces of power plants, the volatile matter is vaporized and carbon is burned off, whereas most of the minerallic matter such as quartz, clays, and feldspars are converted to ash. Some ash is discharged from the bottom of the furnace and the rest leaves the furnace as particles suspended in the flue gases. Fly ash refers to the finer fractions of particulate which is separated from the flue gases by electrostatic precipitator, fabric filters or bag houses to prevent air pollution by fine particulate emission.

The gaseous emissions resulting from coal burning contain the primary pollutants such as sulphur oxides (SO_x) and nitrogen oxides (NO_x). The majority of utility companies are adopting systems for controlling the emissions of gaseous pollutants to meet the regulatory limits. In addition to the installation of wet scrubbers, the other alternatives for pollution control include the use of low sulphur coal, coal cleaning, scrubbing via regeneration process, and dry scrubbing.

11.1.1. SO_x Emissions

On a worldwide basis, man made sources contribute to about 60% of all the sulphur bearing emissions into the atmosphere. The remainder comes from ocean spray, decay of organic matter and volcanoes. In industrialized areas, human activity may contribute up to 90% of all sulphur emissions. Of this 82% is derived from combustion of fossil fuels. 56% of the sulphur bearing emissions results from coal burning and 26% from combustion of petroleum products. About 90% of the sulphur released from man made sources is emitted as sulphur dioxide. In the atmosphere SO_2 undergoes chemical conversion and reaction with other atmospheric species to form a variety of compounds, the most important being sulphuric acid and ammonium sulphate. These compounds in the form of acid rain have direct detrimental effect on human health. Dry deposited SO_2 has deleterious effect on building materials. To overcome this pollution and undesirable effects on human health and materials, many utility companies using coal burning power generation are installing wet scrubbers to reduce sulphur oxide emissions.

11.1.2. NO_x Emissions

In addition to SO_2, the other important contaminant in exhaust gases resulting from coal burning is nitric oxide. It is subsequently oxidized in the atmosphere to nitrogen dioxide, NO_2, a more toxic substance. These two gases are generally collectively referred to as nitrogen oxides, NO_x. On a global basis, fossil fuel combustion accounts for about 40% of NO_x emission, whereas in North America, fossil fuel combustion appears to contribute 85% of the NO_x emissions to the atmosphere. In the United States, coal burning produces about 32% of NO_x emissions. Low NO_x burners are being employed by many coal burning power stations to reduce NO_x emissions to the atmosphere.

11.1.3. PAH Emissions

Recently, another issue has arisen concerning the presence of polycyclic aromatic hydrocarbons (PAH) or also called poly-nuclear aromatic hydrocarbons (PNAs) in emissions from coal burning operations. However, the studies have demonstrated that modern well operated coal-fired power plants produce exhaust gases containing quantities of PAH that are below levels of environmental concern.

11.1.4. Coal Combustion Solid Waste

The solid emissions from coal burning power plants comprising of fly ash, bottom ash and/or boiler slag, and FGD or scrubber waste are effectively utilized and or disposed of in an environmentally satisfactory manner to avoid pollution of land, as well as surface, and ground water. Many utilization schemes of coal combustion by-products, particularly fly ash have been discussed in detail in previous chapters. Systems for environmentally sound and economically efficient utilizations as well as disposal of FGD wastes are being developed for commercial production of marketable products like synthetic gypsum and light weight aggregates (Poz-O-Lite).

11.1.5. Radio Activity

Last but equally important source of pollution is from the radio activity of some of the coal ashes. The senior author is aware of the studies in Poland which suggested that radio activity may be of concern in some specific cases, where the coal itself contains high amount of radio active ingredients. During a visit to Poland it was revealed by some scientists that the incidence of cancer was much more in particular buildings constructed with concrete blocks containing radio active fly ash. Many studies have been conducted in the Netherlands and the United States to determine if radon emissions are a problem in the fly ash concrete. The test data suggest no such problem for the fly ashes tested so far. Apparently radio activity is source specific and is not common to all the fly ashes.

11.2.   CONSTRAINTS ON PRODUCTIVE FLY ASH UTILIZATION

The vast potential for economic utilization of coal combustion by-products including fly ash has been critically reviewed in the present work. The use of coal-ash residue depends on its physical and chemical properties and the variability in quality as well as in quantity. Based on the present state of knowledge, the following factors are identified as the major constraints on potential utilization of coal-ash and other coal combustion by-products.

- Technical appropriateness for their use in field applications.
- Comparative market price with respect to other equivalent natural or traditional materials.
- Environmental concerns such as dusting problems during handling and leachability of soluble salts including trace heavy metals.
- Attitude of potential user industries as well as the general public.
- Transportation and storage facilities. If user markets are far away from the plant, the transportation costs escalate and it may be more economical to dispose of the by-products in landfills than trying to use them. Lack of storage facilities at the plant site also restrains their utilization potential.
- Nonavailability of consistent supply of desired quality coal ash.
- Lack of governmental support in the form of subsidies and incentive for mandatory or regulatory use of fly ash as construction material for public or government funded project.
- Lack of quality control and quality assurance of fly ash at production, storage and supply sources.
- Lack of initiatives in marketing of the by-products by the producers/vendors.
- Trained personnel, time, plant facilities and money required in incorporating the use of the by-products in conventional materials especially for quality assurance of the resultant product or construction material.
- Need for the use of additives or admixtures for obtaining satisfactory and acceptable mortar/concrete mixes for field applications.

- Natural tendency to avoid the risk involved with the use of new experimental material as alternative to the tried and established construction materials. Lack of confidence in using the by-products in construction.
- Erratic air entraining agent demand with fly ash in concrete.
- Reduced resistance to scaling by de-icing chemicals in fly ash concrete.
- Potential variability of quantity of fly ash even from the same source.
- Effects of mix proportioning and laboratory testing when fly ash is incorporated as an additional ingredient of concrete.
- Regulatory and institutional constraints involving requirements, standards, guidelines, specifications, policies and attitudes which discourage unhindered and expanded utilization in engineering applications.

Bibliography

AASHTO M295 (1995). Specification for Fly Ash and Raw or Calcified Natural Pozzolans for Use as a Mineral Admixture in Portland Cement Concrete, FHWA, TRB, Washington, DC.

Abdun Nur, E.A. (1961). Fly Ash in Concrete, an Evaluation. Highway Research Bulletin 284 pp.

Abrams, D.A. (1918). Design of Concrete Mixtures. Bulletin No. 1 Structural Materials Research Laboratory, Lewis Institute, Chicago.

ACI Committee 207 (1980). Roller Compacted Concrete. ACI Journal, Vol. 77, No. 4. Report No. ACI 207.5R-80, pp. 215–236.

ACI Committee 211.1.81 (1984). Standard Practice for Selecting Proportions for Normal, Heavy Weight and Mass Concrete. ACI Manual of Concrete Practice 211.1.81.

ACI Committee 226 (1987). The Use of Fly Ash in Concrete. ACI Materials Journal, Vol. 84, No. 5, pp. 381–409, September–October.

ACI Committee 226-3R-87 (1987). Use of Fly Ash in Concrete. Report by ACI Committee 226, pp. 1–10.

ACI Standard 207.5R-80 (1985). Roller Compacted Concrete. ACI Manual of Concrete Practice, Detroit, Michigan.

ACI Standard 306R-78 (1983). Cold Weather Concreting, American Concrete Institute, Detroit, Michigan.

Akkad-Salam, M. (1992). Properties of Concrete Containing High Amounts of Alberta Fly Ashes, MSc Thesis, University of Calgary, Calgary, Alberta, Canada, January.

Alasali, M.M., and Malhotra, V.M. (1990). Role of High Volume Fly Ash Concrete in Controlling Expansion Due to Alkali–Aggregate Reactions. CANMET International Workshop of Fly Ash in Concrete, Calgary, Canada.

American Petroleum Institute (1985). Standards 10A. Specifications for Oil Well Cements and Cement Additives.

Andrade, C. (1986). Effect of Fly Ash in Concrete on the Corrosion of Steel Reinforcement. ACI, SP-91, pp. 609–620.

Anon. (1951). Relationship of Fly Ash and Corrosion. ACI Journal Vol. 47, pp. 747.

Asrow, S. (1976). Fly Ash Usage in Large Commercial Office Buildings. Proceedings, 4th International Ash Use Symposium. Energy Research and Development Administration, Report No. MERC/SP-76/4, Morgantown, West Virginia, pp. 508–517.

ASTM (1962). Committee C-9 on Concrete and Concrete Aggregates, Cooperative Tests of Fly Ash as an Admixture in Portland Cement Concrete. ASTM Proceedings 62, pp. 314–342.

ASTM (1993). Annual Book of ASTM Standards, Section 4, Construction, Vol. 04-02. Concrete and Mineral Aggregates. Philadelphia, USA.

ASTM C311-93 (1993). Standard Test Methods for Sampling and Testing Fly Ash or Natural Pozzolans for Use as a Mineral Admixture in Portland Cement Concrete. Section 4, Vol. 04.02 Annual Book of ASTM Standards, Philadelphia, USA.

ASTM C618-93 (1993). Specification for Fly Ash and Raw of Calcined Natural Pozzolan for Use a Mineral Admixture in Portland Cement Concrete. Philadelphia, USA.

ASTM C90 (1993). Specification for Hollow Load Bearing Concrete Masonry Units. Philadelphia, USA.

Backes, H.P. (1986). Carbonic Acid Corrosion of Mortars Containing Fly Ash. ACI SP-91, pp. 621–634.

Bamforth, P.B. (1980). In-Situ Measurement of the Effect of Partial Portland Cement Replacement Using Either Fly Ash or Ground Granulated Blast Furnace Slag on the Performance of Mass Concrete. Proceedings, Institution of Civil Engineering, Vol. 69, pp. 777–800.

Baron, J., Bollotte, B., and Clergue, C. (1994). Fly Ash Replacement of Cement Threshold 'Values of Cement Content in Relation to Concrete Durability. P.K. Mehta Symposium on Durability of Concrete, Nice, France, pp. 21–34, May.

Barrer, R.M. (1951). Diffusion in and through Solids. Cambridge University Press, Cambridge, England.

Beal, D.L., and Brantz, H.L. (1992). Assessment of the Durability Characteristics of Triple Blended Cementitious Materials. Fourth CANMET/ACI International Conference on Fly Ash, Silica Fume, Slag and Natural Pozzolans in Concrete. Istanbul, Turkey. ACI SP-132, pp. 1041–1054.

Berry, E.E., and Malhotra, V.M. (1982). Fly Ash for Use in Concrete: A Critical Review. ACI Journal, Vol. 2, No. 3, pp. 59–73.

Berry, E.E., and Malhotra, V.M. (1986). Fly Ash in Concrete. Publication SP 85-3, CANMET, Energy, Mines and Resources, Canada, 178 pp.

Berry, E.E., and Malhotra, V.M. (1987). Fly Ash in Concrete, Chapter 2, Supplementary Cementing Materials. Publication, Supplementary Cementing Materials for Concrete, CANMET, Energy, Mines and Resources, Canada. Edited by V.M. Malhotra, pp. 37–163.

Berry, E.E., Hemmings, R.T., and Cornelius, B.J. (1990). Mechanisms of Hydration Reaction in High Volume Fly Ash Pastes and Mortars. CANMET, International Workshop on Fly Ash in Concrete, Calgary, Canada.

Berry, E.E., Zhang, M.H., Hemmings, R.T., Cornelius, B.J., and Golden, D.M. (1993). Hydration in High Volume Fly Ash Concrete Binders, Proceedings, 10th International Ash Use Symposium, ACAA, EPRI TR-101774, Vol. 2, pp. 51-1 to 51-13, January.

Berube, M.A., and Duchesne, J. (1992). Evaluation of Testing Methods Used for Assessing the Effectiveness of Mineral Admixtures in Suppressing Expansion due to Alkali Aggregate Reaction. ACI SP-132, pp. 549–576.

Bickley, J.A., and Payne (1979). High-Strength Cast-in-Place Concrete in Major Structures in Ontario. Present at ACI Annual Convention. Milwaukee, March.

Biczok, I. (1964). Concrete Corrosion and Concrete Protection. Hungarian Academy of Sciences, Budapest.

Bildeau, A., and Malhotra, V.M. (1992). Concrete Incorporating High Volumes of ASTM Class F Fly Ashes: Mechanical Properties and Resistance of Deicing Salt Scaling and to Chloride Ion Penetration. ACI SP-132, pp. 319–350.

Blick, R.L., Peterson, C.F., and Winter, M.E. (1974). Proportioning and Controlling High Strength Concrete. ACI SP-46, No. 9, pp. 141–163.

Branca, C., Fratesi, R., Moriconi, G., and Simoncini, S. (1992). Influence of Fly Ash on Concrete Carbonation and Rebar Corrosion. ACI SP-132, pp. 245–256.

Brink, R.H., and Halstead, W.J. (1956). Studies Relating to the Testing of Fly Ash for Use in Concrete. Proceedings ASTM 56, pp. 1161–1206.

British Standards Institution, BS3892 (1992). Part I. Specifications for Pulverized Fuel Ash as a Cementitious Component for Use with Portland Cement. BSI, London.

British Standards Institution, BS6610 (1985). Specifications for Pozzolanic Cement and Pulverized Fuel Ash as Pozzolan. BSI, London.

Brown, J.H. (1982). The Strength and Workability of Concrete with PFA Substitution. Proceedings, International Symposium on the Use of PFA in Concrete. University of Leeds, England, pp. 151–161, April.

Brown, P.W., Clifton, J.R., Frohnsdorrf, G., and Berger, R.L. (1976). Limitations to Fly Ash Use in Blended Cements. Proceedings, 4th International Ash Utilization Symposium, St. Louis, ERDA MERC/SP-36/4, pp. 518–529.

Burns, J.S., Guarnaschelli, C., and McAskill, J. (1982). No Controlling the Effect of Carbon in Fly Ash on Air Entrainment. Proceedings, Sixth International Symposium on Fly Ash Utilization, Reno, Nevada, DOE/METC 82-52, pp. 294–313, March.

Buttler, F.G., Decter, M.H., and Smith, G.R. (1983). Studies on the Desiccation and Carbonation of Systems Containing Portland Cement and Fly Ash. ACI SP-79, Vol. 1, pp. 367–383.

Canadian Standards Association CSA/CAN 3-A23.5-M86 (1986). Supplementary Cementing Materials, Rexdale, Ontario, Canada.

CANMET (1985). Compilation of Abstracts of Papers from Recent International Conferences and Symposia on Fly Ash in Concrete. Division Report MRP/MSL 85-2.

CANMET/ACI. International Conferences on Fly Ash, Silica, Fume Slag and Natural Pozzolans in Concrete. ACI Special Publications SP-79 (1983), SP-91 (1986), SP-114 (1989), SP-132 (1992), and SP-153 (1995). Edited by V.M. Malhotra. Published by ACI, Detroit, Michigan.

Cannon, R.W. (1968). Proportioning Fly Ash Concrete Mixes for Strengths and Economy. ACI Journal, Vol. 65, pp. 969–979.

Cannon, R.W. (1972). Concrete Dam Construction Using Earth Compaction Methods. Economical Constructional of Concrete Dams. ASCE, New York, pp. 143–152.

Carette, G.G., and Malhotra, V.M. (1983). Early-Age Strength Development of Concrete Incorporating Fly Ash and Silica Fume. ACI SP-79, pp. 765–784.

Carette, G.G., and Malhotra, V.M. (1984). Characterization of Canadian Fly Ashes and their Performance in Concrete. Division Report MRP/MSL 84-137 (OP&J), CANMET, Energy, Mines and Resources, Canada.

Carette, G.G., and Malhotra, V.M. (1987). Characterization of Canadian Fly Ashes and Their Relative Performance in Concrete. Canadian Journal of Civil Engineering, Vol. 14, pp. 267–282.

Carette, G.G., Painter, K.E., and Malhotra, V.M. (1982). Sustained High Temperature Effect on Concretes Made with Normal Portland Cement, Normal Portland Cement and Slag or Normal Portland Cement and Fly Ash. Concrete International Vol. 4, pp. 41–51, July.

Clarke, Lee B. (1993). Utilization Options for Coal-Use Residues — An International Overview. Proceedings, 10th International Ash Use Symposium ACAA, EPRI TR-101774, Vol. 1, pp. 66-1 to 66-14, January.

Compton, F.R., and McInnis, C. (1952). Field Trial of Fly Ash Concrete Ontario Hydro Research News, pp. 18–21, January–March.

Condry, L.Z. (1976). Recovery of Alumina from Coal Refuse — An Annotate Bibliography, Coal Research Bureau Report No. 130, July.

Cook, J.E. (1982). Research and Application of High Strength Concrete Using Class C Fly Ash. Concrete International Vol. 4, July pp. 72–80.

Crow, R.D., and Dunstan, E.R. (1981). Properties of Fly Ash Concrete. Proceedings, Symposium on Fly Ash Incorporation in Hydrated Cement Systems. Edited by Sidney Diamond. Material Research Society, Boston, pp. 214–225.

Daniels, Lee, Stewart, Barry, and Jackson, Meral. (1993). Utilization of Fly Ash to Prevent Acid Mine Drainage from Coal Refuse. Proceedings, 10th International Ash Use Symposium, ACAA, EPRI TR-101774, pp. 22-1 to 22-13, January.

Davis, R.E. (1954). Pozzolanic Materials with Special Reference to their Use in Concrete Pipes. Technical Memo, American Concrete Pipe Association.

Davis, R.E., Carlson, R.W., Kelly, J.W., and David, H.E. (1937). Properties of Cements and Concrete Containing Fly Ash. ACI Journal, Vol. 33, pp. 577–612.

Dhir, R.K., and Jones, M.R. (1990). Influence of PFA on Proportion of Free Chlorides in Salt Contaminated Concrete. Proceedings, International Conference on Corrosion of Reinforcement in Concrete, Wishaw. Edited by C.L. Page, K.W.L. Treadway, and P.B. Bamforth. Elsevier Applied Science, pp. 227–236.

Dhir, R.K., and Jones, M.R. (1993). PFA Concrete: Exposure Temperature Effects on Chloride Diffusion. Cement and Concrete Research, Vol. 23, No. 25, pp. 1105–1114.

Dhir, R.K., Hubbard, F.H., Munday, J.G.L., Jones, M.R., and Duerden, S.L. (1988). Contribution of PFA in Concrete Workability and Strength Development. Cement and Concrete Research, Vol. 18, No. 2, pp. 227–289.

Dhir, R.K., Jones, M.R., and Ahmed, H.E.H. (1990). Determination of Total and Soluble Chlorides in Concrete. Cement and Concrete Research, Vol. 20, No. 4, pp. 579–590.

Dhir, R.K., Jones, M.R., and Seneviratne, A.G.M. (1991). Diffusion of Chloride Ions in Concretes, Influence of PFA Quality. Cement and Concrete Research, Vol. 21, No. 6, pp. 1092–1102.

Dhir, R.K., Jones, M.R., and McCarthy, M.J. (1992). PFA Concrete: Carbonation Induced Reinforcement Corrosion Rates. Proceedings, Institution of Civil Engineers, England, Vol. 96, pp. 335–342, August.

Dhir, R.K., Jones, M.R., and McCarthy, M.J. (1993). PFA Concrete: Chloride Ingress and Corrosion in Carbonated Cover. Proceedings of the Institution of Civil Engineers, England, Structures and Buildings, Vol. 99, No. 2, pp. 167–172, May.

Diamond, S. (1981). Effects of Two Danish Fly Ashes on Alkali-Contents of Cement–Fly Ash Pastes. Cement and Concrete Research, Vol. 11, pp. 383–394.

Diamond, S. (1985). Selection and Use of Fly Ash for Highway Concrete. GHWA/IN/JHRP-85/8. Joint Highway Research Project, Purdue University, West Lafayette, Indiana, September.

Diamond, S., Barneyback. R.S. Jr., and Strubble, L.J. (1981). On the Physics and Chemistry of Alkali–Silica Reactions. Proceedings, 5th International Conference on Alkali–Aggregate Reaction in Concrete, Cape Town, South Africa, March–April.

Dikeou, J.T. (1970). Fly Ash Increases Resistance of Concrete to Sulphate Attack. Water Resources Technical Publications Research Report 23, USBR.

Docter, Bruce A., and Manz, O.E. (1993). Economic Feasibility of Producing Mineral Wool Fibres at a Coal Generation Power Plant. Proceedings, 10th International Ash Use Symposium, ACAA, EPRI, TR-101774, pp. 81-1 to 81-13, January.

Dodson, V.H. (1981). The Effect of Fly Ash on the Setting Time of Concrete. Chemical or Physical. Proceedings, Symposium on Fly Ash Incorporation in Hydrated Cement Systems. Edited by Sidney Diamond. Materials Research Society, Boston, pp. 166–171.

Dodson, V.H. (1985). A New Approach to the Measurement of the Strength Contribution of Fly Ash to Concrete. W.R. Grace, Co. Presented at Fly Ash Workshop, Sponsored by FHWA and Pennsylvania Department of Transportation, Harrisburg, Pennsylvania, April.

Dolen, T.P. (1990a). Performance of Fly Ash in Roller Compacted Concrete at Upper Still Water Dam. CANMET, International Workshop on Fly Ash in Concrete, Calgary, Alberta, Canada.

Dolen, T.P. (1990b). The Use of Fly Ash Concrete for Upper Still Water Dam Roller Compacted Concrete. CANMET, International Workshop on Fly Ash in Concrete, Calgary, Alberta, Canada.

Duncan, M.A.G., Swenson, E.G., Gillott, J.E., and Foran, M. (1973). Alkali–Aggregate Reaction in Nova Scotia. Summary of Five Year Study. Cement and Concrete Research, Vol. 3, pp. 55–69.

Dunstan, E.R. (1976). Performance of Lignite and Sub-Bituminous Fly Ash in Concrete: A Progress Report. Report REC-ERC-76-1, USBR.

Dunstan, E.R. (1980). A Possible Method for Identifying Fly Ashes that will Improve the Sulphate Resistance of Concretes. ASTM Cement, Concrete and Aggregate, Vol. 2, pp. 20–30.

Dunstan, E.R. (1981). The Effect of Fly Ash on Concrete Alkali–Aggregate Reaction. ASTM Cement, Concrete and Aggregate, Vol. 3, pp. 101–104.

Dunstan, M.R.H. (1982). The Use of High Fly Ash Content in Roads. Proceedings, International Symposium on the Use of PFA in Concrete, University of Leeds, England, pp. 277–289, April.

Dunstan, M.R.H. (1983). Development of High Fly Ash Content in Concrete. Proceedings, Institution of Civil Engineers, London, Part I, Vol. 74, pp. 495–513.

Dunstan, M.R.H. (1986). Fly Ash as the Fourth Ingredient in Concrete Mixtures. Proceedings, 2nd International Conference on Fly Ash, Silica Fume, Slag and Natural Pozzolans in Concrete. ACI SP-91, pp. 171–200.

Dunstan, M.R.H., Thomas, M.D.A., Cripwell, J.B., and Harrison, D.J. (1992). Investigation into the Long Term In-Situ Performance of High Fly Ash Content Concrete used for Structural Applications. ACI SP-132, pp. 1–20.

Edgar, T.F. (1983). Coal Proceedings and Pollution Control. Gulf Publishing Company, Book Division, Houston, Texas.

Electric Power Research Institute (1984). Coal Combustion By-Products Utilization Manual, Vol. 1: Evaluating the Utilization Option, CS-3122, Vol. 12. Annotated Bibliography. EPRI, Palo Alto, California, February.

Electric Power Research Institute (1987). Classification of Fly Ash for Use in Cement and Concrete. CS-5116, Project 2422-10.

Electric Power Research Institute (1993). Proceedings: Tenth International Ash Use Symposium. EPRI TR-101774, Vols. 1 and 2. Orlando, Florida.

Electric Power Research Institute (1995). Proceedings: 11th International Symposium on Use and Management of Coal Combustion By-Products (CCBs). EPRI TR-104657, Vols. 1 and 2. Orlando, Florida.

Elfert, R.J. (1973). Bureau of Reclamation Experiences with Fly Ash and Other Pozzolans in Concrete. Proceedings, 3rd International Ash Utilization Symposium, Pittsburgh Informational Circular, IC 8640, U.S. Bureau of Mines, pp. 80–93.

Elini, Riccardo, Amodie, and Filippo. (1993). Conditioning of Waste Water Sludges by Coal Ash. Proceedings, 10th International Ash Use Symposium, American Coal Ash Association. EPRI TR-101774, pp. 6-1 to 6-11, January.

Environmental Fact Sheet (1992). EPA Guide-Line for Purchasing Cement and Concrete Containing Fly Ash, EPA/530-SW-91-086, 2 pp., January.

Federal Register (1983). Vol. 48, No. 20. Part IV Environmental Protection Agency. Cement and Concrete Containing Fly Ash. Guideline for Federal Procurement, Washington, DC.

FHWA (1986). Fly Ash Facts for Engineers. FHWA Report No. DP 5908 Demonstration Projects Program. Federal Highway Administration, Washington, DC, 47 pp., July.

Fouilloux, P. (1953). French Patent 1,036,771.

Frigione, G., Ferrari, F., and Lanzillotta, B. (1993a). Concretes with High Fractionated Fly Ash Content — Influence of C_3A Content of Portland Cement, Proceedings, 10th International Ash Use Symposium, American Coal Ash Association. EPRI TR-101774, Vol. 2, pp. 46-1 to 46-9, January.

Frigione, G., Lanzillotta, B., Ferrari, F., and Cirillo, G. (1993b). Fly Ash as Basic Raw Material in the Manufacture of Bricks. Proceedings, 10th International Ash Use Symposium, American Coal Ash Association. EPRI TR-101774, Vol. 2, pp. 80-1 to 80-15, January.

Gahlot, P.S., and Lohtia, R.P. (1972). Use of Fly Ash in Mortars. Journal of the Institution of Engineers (India), Vol. 53, C-12, pp. 70–74, November.

Gaynor, R.D., and Mallarky, J.I. (1995). Survey — Use of Fly Ash in Ready Mixed Concrete. National Ready Mixed Concrete Association Silver Spring, Maryland.

Gaze, M.E., and Nixon, P.J. (1983). The Effect of PFA upon Alkali–Aggregate Reaction. Magazine of Concrete Research, Vol. 35, No. 123, pp. 107–110.

Gebauer, J. (1982). Some Observations on the Carbonation of Fly Ash Concrete. Silicate Industrials, Vol. 6, pp. 155–159.

Gebler, S.H., and Klieger, P. (1983). Effect of Fly Ash on the Air Void Stability of Concrete. Proceedings, 1st International Conference on the Use of Fly Ash, Silica Fume, Slag and Other Mineral By-Products in Concrete. ACI SP-79, pp. 103–142, Detroit, Michigan.

Gebler, S.H., and Klieger, P. (1986a). Effect of Fly Ash on Physical Properties of Concrete. Proceedings, 2nd International Conference on Fly Ash, Silica Fume, Slag and Natural Pozzolans in Concrete. ACI SP-91, Vol. 1, pp. 1–50, Detroit, Michigan.

Gebler, S.H., and Klieger, P. (1986b). Effect of Fly Ash on the Durability of Air-Entrained Concrete. Proceedings, 2nd International Conference on Fly Ash, Silica Fume, Slag and Natural Pozzolans in Concrete. ACI SP-91, Vol. 1, pp. 483–520, Detroit, Michigan.

Geiker, M., and Thaulow, N. (1992). The Mitigating Effect of Pozzolans on Alkali Silica Reactions. ACI SP-132, Vol. 1, pp. 533–548, Detroit, Michigan.

Geisler, J., and Lang, E. (1994). Long Term Durability of Non-Air Entrained Concrete Structures Exposed to Marine Environments and Freezing and Thawing Cycles. 3rd CANMET/ ACI International Conference on Durability of Concrete, Nice, France, pp. 715–737.

Ghosh, R.S. (1976). Proportioning Concrete Mixes Incorporating Fly Ash. Canadian Journal of Civil Engineering Vol. 3, pp. 168–182.

Ghosh, R.S., and Timusk, J. (1981). Creep of Fly Ash Concrete. ACI Journal, Vol. 78, No. 5, pp. 351–357.

Giergiczny, Z. (1992). The Properties of Cements Containing Fly Ash Together with Other Admixtures. ACI SP-132, pp. 457–470, Detroit, Michigan.

Gifford, P.M., Langan, B.W., Day, R.L., Joshi, R.C., and Ward, M.A. (1987). A Study of Fly Ash Concrete in Curb and Gutter Construction Under Various Laboratory and Field Curing Regimes. Canadian Journal of Civil Engineering, Vol. 14, No. 5, pp. 614–620, October.

Gifford, P.M., Langan, B.W., and Ward, M.A. (1990). A Study of the Use of Fly Ash in Concrete at Calgary, Canada.

Gilliland, J.L. (1951). Relationship of Fly Ash and Corrosion. ACI Journal, Vol. 47, p. 397.

Gillott, J.E. (1980). Properties of Aggregate Affecting Concrete in North America, G.Q. Eng. Geol., Vol. 13, London, pp. 289–303.

Golden, D.M. (1990). EPRI, R&D Program in High Volume Fly Ash Concrete. CANMET, International Workshop of Fly Ash in Concrete, Calgary, Canada.

Goma, F. (1992). Concrete Incorporating a High Volume of ASTM Class C Fly Ash with High Sulphate Content. ACI SP-132, pp. 403–418, Detroit, Michigan.

Guiguan, S., Luoshu, G., and Haimin, W. (1986). Concrete Made With Calcium Enriched Fly Ash. Proceedings, 2nd International Conference on Fly Ash, Silica Fume, Slag and Natural Pozzolans in Concrete. ACI SP-91, pp. 387–412, Detroit, Michigan.

Hague, M.N., Day. R.L., and Langan, B.W. (1988a). Realistic Strength of Air Entrained Concretes With and Without Fly Ash. ACI Materials Journal, pp. 241–247, July–August.

Hague, M.N., Langan, B.W., and Ward, M.A. (1988b). High Fly Ash Concretes. Journal of ACI, Vol. 8, No. 1, pp. 54–60.

Halow, J.S., and Covey, J.N. (Editors) (1982). The Challenge of Change — Sixty International Ash Utilization Symposium Proceedings, DOE/MWRX/8-52, Vols. 2 and 3. Morgantown and Energy Technology Center, U.S. Department of Energy and National Ash Association, Morgantown, West Virginia, July.

Halstead, W.J. (1981). Quality Control of Highway Concrete Containing Fly Ash. VHTRC 81-R38. Virginia Highway and Transportation Research Council, Charlottesville, Virginia, February.

Hansen, K.P., and Reinhardt, W.P. (1991). Roller Compacted Concrete Dams, McGraw-Hill Inc.

Helmuth, R.A. (1986). Water Reducing Properties of Fly Ash in Cement Pastes, Mortars, and Concretes, Causes and Test Methods. Proceedings, 2nd International Conference on Fly Ash, Silica Fume, Slag and Natural Pozzolans in Concrete. ACI SP-91, Vol. 2, pp. 723–740, Detroit, Michigan.

Ho, D.W.S., and Lewis, R.K. (1983). Carbonation of Concrete Incorporating Fly Ash or a Chemical Admixture. ACI SP-79, pp. 333–346.

Hobbs, D.W. (1981). The Alkali–Silica Reaction: A Model for Predicting Expansion in Mortar. Magazine of Concrete Research, Vol. 33, pp. 208–220.

Hobbs, D.W. (1982). Influence of Pulverized Fuel Ash and Granulated Blast Furnace Slag Upon Expansion Caused by the Alkali Silica Reaction. Magazine of Concrete Research, Vol. 34, pp. 83–93.

Hobbs, D.W. (1983a). Influence of Fly Ash on the Workability and Early Strength of Concrete. Proceedings, 1st International Conference on the Use of Fly Ash, Silica Fume, Slag and other Mineral By-Products in Concrete. ACI SP-79, pp. 289–306, Detroit, Michigan.

Hobbs, D.W. (1983b). Possible Influence of Small Additions of PFA, GBFS and Limestone Flour Upon Expansion of Concrete. Magazine of Concrete Research, Vol. 35, pp. 35–55.

Hoff, C.G., and Buck A.D. (1983). Consideration in the Prevention of Damage to Concrete Frozen at Early Ages, Technical Paper Title No. 80-35. ACI Journal, pp. 371–375.

Idorn, G.M. (1991). Concrete Durability and Resource Economy Concrete International, Vol. 13, No. 7, pp. 18–23.

Ivanhov, Ya., and Zachavrieva, S. (1982). Influence of Fly Ash on Rheology of Fresh Concrete. Proceedings, International Symposium on the Use of PFA in Concrete, University of Leeds, England. Edited by J.A. Cabyera and A.R. Cusens, pp. 133–141.

Jawed, I., and Skalny, J. (1977). Alkalies in Concrete: A Review. Cement Concrete Resources, Vol. 7, pp. 719–730.

Johnson, G.E., Kunka, L.M., Forney, A.J., and Field, J.H. (1966). The Use of Coal and Modified Coal as Adsorbents for Removing Organic Contaminants from Waste Waters. Bureau of Mines Report of Inv. 6884, 56 pp.

Johnston, C.D. (1994). Deicer Salt Scaling Resistance and Chloride Permeability. Concrete International, pp. 48–55, August.

Jones, M.R., McCarthy, M.J., and Dhir, R.K. (1993). Chloride Resistance Concrete. Proceedings, International Conference Concrete 200, Economic and Durable Construction Through Excellence, Dundee. Edited by R.K. Dhir and M.R. Jones. E&FNspon, pp. 1429–1444, September.

Joshi, R.C. (1968). Experimental Production of Synthetic Fly Ash from Kaolinite. MS Thesis. Iowa State University, Ames.

Joshi, R.C. (1970). Pozzolanic Reactions in Synthetic Fly Ashes. PhD Dissertation, Iowa State University, Ames.

Joshi, R.C. (1979). Sources of Pozzolanic Activity in Fly Ashes — A Critical Review. Proceedings, 5th International Fly Ash Utilization Symposium, Atlanta, Georgia, USA, pp. 610–623.

Joshi, R.C. (1982). Effect of Coarse Fraction (+#325) of Fly Ash on Concrete Properties. Proceedings, Sixth International Symposium on Fly Ash Utilization, Reno, Nevada, pp. 77–85, March.

Joshi, R.C. (1987). Effect of a Sub-Bituminous Fly Ash and its Properties on Sulphate Resistance of Sand Cement Mortars. Journal of Durability of Building Materials, Vol. 4, pp. 271–286.

Joshi, R.C., and Balasundaram, A. (1988). Some Properties of Cement and Cement/Fly Ash Mortars Prepared with Sea

Water. Proceedings, 2nd International Conference on Performance of Concrete in Marine Environment, Sponsored by CANMET/ACI, pp. 173–190, August.

Joshi, R.C., and Lam, D.T. (1987). Sources of Self Hardening Properties in Fly Ashes. Materials Research Proceedings, Vol. 86, MRS Pittsburgh, USA, pp. 183–184.

Joshi, R.C., and Lohtia, R.P. (1993a). Effects of Premature Freezing Temperatures on Compressive Strength, Elasticity and Microstructure of High Volume Fly Ash Concrete. Third Canadian Symposium on Cement and Concrete (CSCC, 93), Ottawa, Canada, August.

Joshi, R.C., and Lohtia, R.P. (1993b). Types and Properties of Fly Ash. Mineral Admixtures in Cement and Concrete. Progress in Cement and Concrete, Vol. 4. Edited by S.L. Sarkar et al. ABI Books Pvt. Ltd, New Delhi, India, pp. 118–157.

Joshi, R.C., and Lohtia, R.P. (1995). Fly Ash Classification System Based on Loss on Ignition (LOI). Proceedings, 11th International Symposium on Use and Management of Coal Combustion By-Products (CCBs). Vol. 2, pp. 61-1 to 61-14.

Joshi, R.C., and Marsh, B.K. (1987). Some Physical, Chemical and Mineralogical Properties of Some Canadian Fly Ashes. Materials Research Society, Proceedings, Vol. 86, MRS. Pittsburgh, USA, pp. 113–126.

Joshi, R.C., and Nagary, T.S. (1983). Generalization of Flow Behaviour of Cement–Fly Ash Pastes and Mortars. Journal of Materials Sciences, ASCE, Vol. 2, No. 3, pp. 128–135, August.

Joshi, R.C., and Natt, G.S. (1983). Roller Compacted High Fly Ash Concrete (Geocrete). ACI SP-79, pp. 347–366.

Joshi, R.C., and Rosauer, E.A. (1973a). Pozzolanic Activity in Synthetic Fly Ashes — Experimental Production and Characterization. American Ceramic Society Bulletin, Vol. 52, No. 5, pp. 456–459, May.

Joshi, R.C., and Rosauer, E.A. (1973b). Pozzolanic Activity in Synthetic Fly Ashes II, Pozzolanic Behaviour, American Ceramic Society Bulletin, Vol. 52, No. 5, pp. 459–463, May.

Joshi, R.C., and Salam, M.A. (1992). High Volume Fly Ash High Strength Concrete. Proceedings, Canada–Japan Seminar on Concrete Technology, University of Calgary, Calgary, Alberta, Canada, August.

Joshi, R.C., and Ward, M.A. (1980). Self Cementitious Fly Ashes Structure and Hydration Mechanism. Proceedings, the International Congress on the Chemistry of Cement. Editions Septima, pp. IV/78–IV/83, Paris.

Joshi, R.C., Day, R.L., Langan, B.W., and Ward, M.A. (1986). Engineering Properties of Concrete Containing High Proportions of Fly Ash and Other Mineral Admixtures. Presented at 2nd International Conference on Use of Fly Ash, Silica fume, Slag and National Pozzolans in Concrete, Madrid, Spain, Sponsored by CANMET, ACI and Spanish Concrete Association.

Joshi, R.C., Salam, M.A., and Wijeweera, H. (1991). Durable and High Strength Concrete with 40% or More Fly Ash in Place of Cement. Alberta Municipal Affairs, Edmonton, Canada. Report No. 0-88654-349-5, 94 pp., May.

Joshi, R.C., Lohtia, R.P., and Salam, M.A. (1993). High Strength Concrete with High Volumes of Canadian Sub-Bituminous Coal Fly Ash. Third International Symposium on Utilization of High Strength Concrete, Lillehammer, Norway, June.

Joshi, R.C. Lohtia, R.P., and Salam, M.A. (1994). Some Durability Related Properties of Concretes Incorporating High Volumes of Sub-Bituminous Coal Fly Ash. Proceedings: 3rd CANMET/ACI International Conference on Durability of Concrete, Nice, France, pp. 447–464.

Joshi, R.C., Lohtia, R.P., and Achari, G. (1995). Fly Ash–Cement Mixtures For Solidification and Detoxification of

Oil/Gas Well Sludges. Proceedings: 74th Annual TRB Meeting, January.

Kalousek, G.L., Porter, L.C., and Benton, E.G. (1972). Concrete for Long Time Service in Sulphate Environment, Cement and Concrete Research, Vol. 2, pp. 79–89.

Kanitakis, I.M. (1981). Permeability of Concrete Containing Pulverized Fuel Ash. Proceedings, 5th International Symposium on Concrete Technology, Nuevo Leon, Mexico, pp. 311–322, March.

Kasai, Y., Matsui, I., Fukushima, U., and Kamohara, H. (1983). Air Permeability of Blended Cement Mortars. Proceedings, 1st International Conference on the Use of Fly Ash, Silica Fume, Slag and Other Mineral By-Products in Concrete. ACI SP-79, pp. 435–451.

Katell, S. (1977). The Potential Economics of the Recovery of Trace Elements in Coal Refuse. Coal Research Bureau Report No. 142, West Virginia University, Morgantown.

Katsiris, N., and Kouzeli-Katsiris, A. (1987). Bound Water Content of Biological Sludges in Relation to Filtration and Dewatering Water Research, Vol. 21.

Kobayashi, K. (1972). Japan Patent 7245,925.

Kobayashi, M. (1979). Utilization of Fly Ash and its Problems in Use in Japan, Japan/U.S. Science Seminar, San Francisco, pp. 61–69, September.

Kokubu, M., and Nagataki, S. (1969). Carbonation of Concrete Correlating with the Corrosion of Reinforcement in Fly Ash Concrete, Proceedings Symposium on Durability of Concrete, Rilem, Final Report, Part II, D71–D79.

Kokubu, M., Muira, I., Takano, S., and Sugiki, R. (1960). Effect of Temperature and Humidity during Curing on Strength of Concrete Containing Fly Ash. Transactions Japan Society of Civil Engineers, No. 7L, Extra Papers (4-3), December.

Kondo, J., Takeda, A., and Mideshima, S. (1958). Effect of Admixtures on Electrolytic Corrosion of Steel Bars in Reinforced Concrete. Journal Japanese Society of Civil Engineers, Vol. 43, pp. 1–8.

Kovac, V., and Ukraincik, V. (1983). Studies in the Use of Fly Ash in Concrete for Water Dam Structures. ACI SP-79, pp. 173–185.

Lai, C.I. (1992). Strength Characteristics of Flowable Mortars Containing Coal Ash. ACI SP-132, pp. 119–134.

Lane, R.O., and Best, G.F. (1982). Properties and Use of Fly Ash in Portland Cement Concrete. Concrete International, Vol. 4, pp. 81–92, July.

Langan, B.W., Joshi, R.C., and Ward M.A. (1983). Strength and Durability of Concretes Containing 50% Portland Cement Replacement by Fly Ash and Other Materials.

Langley, W.S. (1990). Long Term Strength Development and Temperature Rise in Mass Concrete Containing High Volumes of Low Calcium (ASTM Class F) Fly Ash, CANMET, International Workshop of Fly Ash in Concrete, Calgary, Canada.

Langley, W.S., Carette, G.G., and Malhotra, V.M. (1989). Structural Concrete Incorporating High Volumes of ASTM Class F Fly Ash. ACI Materials Journal 86-M48, Vol. 86, No. 5, pp. 507–514.

Larson, T.D. (1953). Effect of Substitutions of Fly Ash for Portland Cement in Air Entrained Concrete. Proceedings: 32nd Annual Meeting Highway Research Board, pp. 328–335.

Larson, T.D. (1964). Air Entrainment and Durability Aspects of Fly Ash Concrete. ASTM Proceedings 64, pp. 866–886.

Larsen, T.J. (1985). Use of Fly Ash in Structural Concrete in Florida. Presented at Fly Ash in Highway Construction Seminar, Atlanta, Georgia, March.

Larsen, T.J., McDaniel, W.H., Brown, R.P., and Sosa, J.L. (1976). Corrosion Inhibiting Properties of Portland and Portland Pozzolan Cement Concrete. Transportation Research Records No. 613, pp. 21–29.

Lawrence, William F. (1969). Mineral Wool Production from Coal Ash — A Progress Report #38, Coal Research Bureau, West Virginia University.

Liu, T.C. (1980). Maintenance and Preservation of Concrete Structures. Report 3. Abrasion/erosion Resistance of Concrete. Technical Report C-78-4, U.S. Army Waterways Experiment Station. 129 pp., July.

Lohtia, R.P. (1972). The Use of Indigenous Admixtures in Cement Concrete. Journal of Cement and Concrete Research, Vol. 12, No. 4, pp. 324–337, January.

Lohtia, R.P. (1991). The Performance of Indigenous Superplasticizers in Cement Mortar and Concrete. Indian Concrete Journal, Bombay, June.

Lohtia, R.P., and Gahlot, P.S. (1973). Design of Fly Ash Concrete Mixes. Journal of Irrigation and Power, Vol. 30, No. 2, New Delhi, India, pp. 171–180, April.

Lohtia, R.P., and Joshi, R.C. (1995). Mineral Admixtures. Concrete Admixtures Handbook: Properties, Science, and Technology. Second Edition. Edited by V.S. Ramachandran. Noyes Publications, Park Ridge, New Jersey, USA, pp. 657–739.

Lohtia, R.P., Nautiyal, B.D., and Jain, O.P. (1976). Creep of Fly Ash Concrete. ACI Journal, Vol. 73, pp. 469–472.

Lohtia, R.P., Nautiyal, B.D., Jaen, K.K., and Jain, O.P. (1977). Compressive Strength of Plain and Fly Ash Concrete by Non Destructive Testing Methods, Journal of the Institution of Engineers (India), Vol. 58a-1, pp. 40–45.

Lohtia, R.P., Joshi, R.C., and Akkad-Salam, M. (1992). Effect of Premature Freezing on Properties of High Volume Fly Ash Concrete Proceedings, Canada–Japan Seminar on Concrete

Technology, University of Calgary, Calgary, Alberta, Canada, August.

Lovewell, C.E., and Washa, G.W. (1958). Proportioning Concrete Mixtures Using Fly Ash. ACI Journal, Vol. 54, pp. 1093–1102.

Mailvagnam, N.P., Bhagrath, R.S., and Shaw, K.L. (1983). Effect of Air Entrainment of Portland Cement Concrete Incorporating Blast Furnace Slag and Fly Ash. ACI SP-79, pp. 519–537.

Malhotra, V.M. (1981). Mechanical Properties and Durability of Superplasticized Semi-Light Weight Concrete. In Developments in the Use of Superplasticizers. ACI SP-68, pp. 283–305.

Malhotra, V.M. (1988). Superplasticized Fly Ash Concrete for Structural Applications. Concrete International, Design and Construction, Vol. 8, No. 12, pp. 28–31.

Malhotra, V.M., and Carette, G.G. (1983). Performance of Concrete Incorporating Limestone Dust as a Partial Replacement of Sand. Report 83-41, CANMET, Energy, Mines and Resources, Canada.

Malhotra, V.M., Carette, G.G., and Bremmer, T.W. (1980). Durability of Concrete Containing Granulated Blast Furnace Slag or Fly Ash or Both in Marine Environment. Report 80-18E, CANMET, EMR, Canada, June.

Malhotra, V.M., Carette, G.G., and Brenner, T.W. (1982). CANMET Investigations Dealing with the Performance of Concrete Containing Supplementary Cementing Materials at Treat Island, Maine.

Malhotra, V.M., Carette, G.G., and Bilodeau, A. (1990a). Fibre-Reinforced High-Volume Fly Ash Shotcrete for Controlling Aggressive Leachate from Exposed Rock Surfaces and Mine Tailings. CANMET, International Workshop on Fly Ash in Concrete, Calgary, Canada.

Malhotra, V.W., Carette, G.G., Bilodeau, A., and Sivasundaram, V. (1990b). Some Aspects of Durability of High-Volume ASTM Class F (Low Calcium) Fly Ash Concrete, Mineral Sciences Laboratories, Division Report MSL 90-20 (OP&J), March.

Malhotra, V.M., Bilodeau, A., Carette, G.G. (1993). Mixture Proportions, Mechanical Properties and Durability Aspects of High Volume Fly Ash Concrete. Proceedings, 10th International Ash Use Symposium, ACAA, EPRI, TR-101774, Vol. 2, pp. 50-1 to 50-7, January.

Malhotra, V.M., Carette, G.G., and Bilodeau, A. (1994). Mechanical Properties and Durability of Polypropylene Fibre-Reinforced, High Volume Fly Ash Concrete for Shotcrete Applications. 3rd CANMET/ACI International Conference on Durability of Concrete, Nice, France, pp. 753–774.

Manmohan, D., and Mehta, P.K. (1981). Influence of Pozzolanic, Slag and Chemical Admixtures on Pore Size Distribution and Permeability of Hardened Cement Pastes. ASTM Cement, Concrete and Aggregates, Vol. 3, No. 1, pp. 63–67.

Manz, O.E. (1980). Utilization of Coal By-Products from Western Coal Combustion in the Manufacture of Mineral Wool and Other Ceramic Materials. Cement and Concrete Research, Vol. 10, pp. 513–520.

Manz, O.E. (1993). Worldwide Production of Coal Ash and Utilization in Concrete and Other Products. Proceedings, Tenth International Ash Use Symposium, Vol. 2, American Coal Ash Association, Washington, DC. EPRI, TR-101774, pp. 64–72.

Manz, O.E., and McCarthy, J.G. (1986). Effectiveness of Western U.S. High-Lime Fly Ash for Use in Concrete. Proceedings, 2nd International Conference on Fly Ash, Silica Fume, Slag and Natural Pozzolans in Concrete. ACI SP-91, pp. 347–366.

Marchard, J., Pigeon, M., Boisvert, J., Isabelle, H.L., and Houdusse, O. (1992). Deicer Salt Scaling Resistance of Roller Compacted Concrete Pavements Containing Fly Ash and Silica Fume. ACI SP-132, pp. 151–178.

Marusin, S.L. (1992). Influence of Fly Ash and Moist Curing Time on Concrete Permeability. ACI SP-132, pp. 257–270.

Mass, G. (1982). Proportioning Mass Concrete and Incorporating Pozzolans Using ACI 211.1 Concrete International, Vol. 4, pp. 45–55.

Mather, B. (1982). Concrete in Sea Water. Concrete International, Vol. 4, pp. 28–34.

Mather, K. (1982). Current Research on Sulphate Resistance at the Waterways Experiment Station. Proceedings of the George Verbeck Symposium on Sulphate Resistance of Concrete. ACI SP-77, pp. 63–74.

Matyszewski, T., Bania, A., and Piasecki, J. (1994). Durability of Concrete in Marine Environment. 3rd CANMET/ACI International Conference on Durability of Concrete, Nice, France, pp. 547–562.

McCarthy, G.J., and Solem, J.K. (1991). X-Ray Diffraction Analysis of Fly Ash II, Results. Advances in X-Ray Analysis, Vol. 34. Edited by C.S. Barrett et al. Plenum Press, New York.

McCarthy, G.J., Johansen, D.M., and Steinwand, S.J. (1988). X-Ray Diffraction Analysis of Fly Ash. Advances in X-Ray Analysis, Vol. 31. Edited by C.S. Barrett et al. Plenum Press, New York.

Mehrotra, A.K., Svreck, W.Y., and Bishnoi, P.R. (1977). Extraction of Metals as their Chlorides from Fly Ash. Proceedings of Coal and Coke Sessions 28th Canadian Chemical Engineering Conference, Halifax, Nova Scotia.

Mehrotra, A.K., Bishnoi, P.R., and Svreck, W.Y. (1979). Metal Recovery from Coal Ash via Chlorination: A Thermodynamic Study. The Canadian Journal of Chemical Engineering, Vol. 57, pp. 225–232, April.

Mehrotra, A.K., Behie, L.A., Bishnoi, P.R., and Svreck, W.Y. (1982). I&EC Process Design and Development. High Temperature Chlorination of Coal Ash in a Fluidized bed. 1. Recovery of Aluminum 2. Recovery of Iron, Silicon, and Titanium, American Chemical Society, Vol. 21, pp. 37–50, January.

Mehta, P.K. (1980). Durability of Concrete in Marine Environment. A Review in Performance of Concrete in Marine Environment. ACI SP-65, pp. 1–20.

Mehta, P.K. (1983). Pozzolanic and Cementitious By-Products as Mineral Admixtures for Concrete — A Critical Review, Proceedings: First International Conference on the Use of Fly Ash, Slag, and Silica Fume in Concrete. Montebello, Canada. ACI SP-79, pp. 1–46.

Mehta, P.K. (1984) Testing and Correlation of Fly Ash Properties with Respect of Pozzolanic Behaviour. CS-3314, EPRI, Palo Alto, California, January.

Mehta, P.K. (1988). Standard Specifications for Mineral Admixtures — An Overview. ACI SP-91, pp. 637–658.

Mehta, P.K. (1989). Pozzolanic and Cementitious By-Products in Concrete — Another Look. Proceedings 3rd International Conference on Fly Ash, Silica Fume, Slag and Natural Pozzolans in Concrete, Norway. ACI SP-114, pp. 1–43.

Mehta, P.K. (1991). Durability of Concrete: Fifty Years of Progress. ACI SP-126. Edited by V.M. Malhotra, pp. 1–32.

Mehta, P.K. (1993). Sulphate Attack on Concrete. A Critical Review. Materials Science of Concrete III. Edited by J. Skalny. American Ceramic Society, pp. 105–130.

Mehta, P.K. (1994a). Symposium on Durability of Concrete, Edited by I.H. Khayad and P.C. Aitcin. Nice, France, pp. 99–118.

Mehta, P.K. (1994b). Symposium on Durability of Concrete, Edited by I.H. Khayad and P.C. Aitcin. Nice, France, pp. 291–335.

Mehta, P.K., and Gjorv, O.E. (1982). Properties of Cement Concrete Containing Fly Ash and Condensed Silica Fume. Cement and Concrete Research, Vol. 12, pp. 587–596.

Meininger, R.C. (1979). Status Report on Effect of Fly Ash on Air Entrainment in Concrete. Series J-153, National Ready Mixed Concrete Association, Silver Springs, Maryland, July.

Meyer, A. (1969). Investigation on the Carbonation of Concrete, Proceedings, 5th International Symposium on Chemistry of Cement, Tokyo, 1968, III-52, pp. 394–401.

Michalski, B. (1991). International Energy Annual 1989. Department of Energy Information Administration, DOE-EIA-0219(89), Table 4, Washington, DC.

Mindess, S., and Young, J.F. (1981). Concrete. Prentice-Hall Inc., Englewood Cliffs, New Jersey. 671 pp.

Minnick, L.T. (1959). Fundamental Characteristics on Pulverized Coal Fly Ashes. Proceedings, Vol. 59, ASTM, Washington, DC.

Mukherjee, P.K., Loughborough, M.T., and Malhotra, V.M. (1981). Development of High Strength Concrete Incorporating a Large Percentage of Fly Ash and Superplasticizers. CANMET, Energy, Mines and Resources, Canada. Project MRP-3.6.0.0.65 (Minerals Research Program), August.

Mukherjee, P.K., Loughborough, M.T., and Malhotra, V.M. (1983). Development of High Strength Concrete Incorporating a Large Percentage of Fly Ash and Superplasticizers. ASTM Cement, Concrete and Aggregates, Vol. 4, pp. 81–86.

Munday, J.G.L., and Dhir, R.K. (1978). Mix Design for Corresponding Strength with Pulverized Fuel Ash as Partial Cement replacement. Proceedings International Conference on Materials of Construction for Developing Countries, Bangkok, pp. 263–273, August.

Munday, J.G.L., Ong, L.T., Wong, L.G., and Dhir, R.K. (1982). Load Independent Movements in OPC/PFA Concrete. Proceedings, International Symposium on the Use of PFA in Concrete. University of Leeds, England. Edited by J.A. Cabreva and A.R. Cusens, pp. 243–246.

Munday, J.G.L., Ong, L.T., and Dhir, R.K. (1983). Mix Proportioning of Concrete with PFA: A Critical Review in Fly Ash Silica Fume, Slag and other Mineral By-Products in Concrete. ACI SP-79, pp. 267–288, Detroit, Michigan.

Murtha, M.J., and Burnet G. (1976). Recovery of Alumina from Coal Ash by High Temperature Chlorination. Proceedings of Iowa Academy of Sciences, Vol. 83, pp. 125–129.

Nagataki, S., and Ohga, H. (1992). Combined Effect of Carbonation and Chloride on Corrosion of Reinforcement in Fly Ash Concrete. ACI SP-132, pp. 227–244.

Nagataki, S., Ohga, H., and Kim, E.K. (1986). Effect of Curing Conditions on the Carbonation of Concrete with Fly Ash and the Corrosion of Reinforcement in Long Term Tests. ACI SP-91, pp. 521–540.

Naik, T.R., Sivasundaram, V., and S.S. Singh. (1991). Use of High Volume Class F Fly Ash for structural Grade Concrete Transportation Research Board, Record No. 1301, Washington, DC, pp. 40–47.

Nakahara, Y., and Yurugu, M. (1991). Laboratory and Field Tests for Methods of Quality Control and Modification of Mix Proportions for Concrete in RCC pavements. Proceedings of Japan Society of Civil Engineers, Vol. 3.

Nasser, K.W., and Al-Manasser, A.A. (1986). Shrinkage and Creep of Concrete Containing 50 Percent Lignite Fly Ash at Different Stress Strength Ratios Proceedings, 2nd International Conference on Fly Ash, Silica Fume, Slag and Natural Pozzolans in Concrete. ACI SP-91, Vol. 1, pp. 433–448.

Nasser, K.W., and Lai, P.S.H. (1992). Resistance of Fly Ash Concrete to Freezing and Thawing. ACI SP-132, pp. 205–226.

Nasser, K.W., and Lohtia, R.P. (1971). Mass Concrete Properties at High Temperatures. ACI Journal, Vol. 68, No. 4, pp. 276–281.

Nasser, K.W., and Lohtia, R.P. (1976). Resistance of Concrete to the Attack of Sodium Sulphate Solution. Proceedings, Rilem Symposium on Durability of Concrete, Prague, Czechoslovakia, September.

Nasser, K.W., and Marzouk, H.M. (1979). Properties of Mass Concrete Containing Fly Ash at High Temperatures. ACI Journal, Vol. 76, No. 4, pp. 537–550.

Nasser, K.W., and Marzouk, H.M. (1983). Properties of Concrete Made with Sulphate Resisting Cement and Fly Ash. Proceedings: First International Conference on the Use of Fly Ash, Silica Fume, Slag and Other Mineral By-Products in Concrete. ACI SP-79, pp. 383–395.

Nasser K.W., and Ojha, R.N. (1990). Properties of Concrete Made with Saskatchewan Fly Ash. CANMET, International Workshop on Fly Ash in Concrete at Calgary, Canada.

National Ready-Mixed Concrete Association (1991). 1987 Survey of Fly Ash Use in Ready-Mixed Concrete, December.

Neville, A.M. (1981). Properties of Concrete, Third Edition. Pitman Publishing Limited, London, 779 pp.

Nishi, S., Ohmori, Y., and Yamamoto, I. (1983). A New Air Entraining Agent Undistributed by the Residual Carbon in Fly Ash. JROCC, Vol. 35, No. 110, pp. 31–37.

Nixon, P.J., and Gaze, M.E. (1983). The Effectiveness of Fly Ashes and Related Blast Furnace Slags in Preventing ADAR. Proceedings, 6th International Conference on Alkalies in Concrete, Copenhagen. Edited by G.M. Doran and S. Rostan, pp. 61–68, June.

Nocum-Wezeik, W. (1994). Carbonation in 50-Year-Old Concrete. Proceedings: 3rd CANMET/ACI International Conference on Durability of Concrete, Nice, France, pp. 289–300.

Nonavinakere, Sujith Kumar, and Reed, Brian E. (1993). Use of Fly Ash to Treat Metal Bearing Waste Streams. Proceedings, 10th International Ash Use Symposium, ACAA, EPRI TR-101774, Vol. 1, pp. 3-1 to 3-15, January.

Oberholster, R.E., and Westra, W.B. (1981). The Effectiveness of Mineral Admixtures in Reducing Expansion Due to Alkali–Aggregate Reaction with Malmesbury Group Aggregate. Proceedings, 5th International Conference on Alkali–Aggregate Reaction in Concrete, Cape Town, South Africa, Paper S252/31, March–April.

Ohga, H., and Nagataki, S. (1992). Effect of Fly Ash on Alkali–Aggregate Reaction in Marine Environment. ACI SP-132, Vol. 1, pp. 577–590.

Owens, P.L. (1979). Fly Ash and Its Usage in Concrete, Concrete Society Journal, England, Vol. 13, No. 7, pp. 21–26.

Oztekin, E. (1988). Accelerated Strength Test Results with Pozzolan Cement Concrete. Proceedings: 2nd International Conference on Fly Ash, Silica Fume, Slag and Natural Pozzolans in Concrete. ACI SP-91, pp. 231–248.

Paillere, A.M., Raverdy, M., and Grimaldi, G. (1986). Carbonation of Concrete with Low Calcium Fly Ash and Granulated Blast Furnace Slag — Influence of Air Entraining Agents and Freezing and Thawing Cycles. ACI SP-91, pp. 541–562.

Paillere, A.M., Platret, G., Roussel, P., and Gawsewitch, J. (1992). Influence of Mechanical Strength and Curing Methods of Sea Water Durability of Mortars Containing Fly Ashes and Slag. ACI SP-132, pp. 179–204.

Panday, K.K., Parsad, G., and Singh, V.N. (1985). Copper (II) Removal from Aqueous Solutions by Fly Ash. Water Research, Vol. 19, No. 7, pp. 869–873.

Paprocki, A. (1970). The Inhibitory Effect of Fly Ash with Respect to Corrosion of Steel in Concrete. Proceedings, 2nd International Ash Utilization Symposium. U.S. Bureau of Mines, Pittsburgh, PA, pp. 17–23.

Papayianni, J. (1992). Performance of a High Calcium Fly Ash in Roller Compacted Concrete. ACI SP-132, pp. 367–386.

PCA (1987). Concrete Information, Structural Design of Roller-Compacted Concrete for Industrial Pavements. PCA, Skokie, Illinois, pp. 1–8.

PCA (1992). RCC Newsletter. Roller Compacted Concrete Design and Construction, Vol. 8, No. 1. Skokie, Illinois, Spring/ Summer.

PCA (1993). RCC Newsletter, Roller Compacted Concrete Design and Construction, Vol. 9, No. 1. Skokie, Illinois, Spring/ Summer.

PCA Publications (1991). Autoclaved Cellular Concrete — The Building Material of the 21st Century. Concrete Technology Today, Vol. 12, No. 2, July.

PCA Publications (1992). Mobil Demonstration Plant Will Produce Fly Ash Based Cellular Concrete. Concrete Technology Today, Vol. 13, No. 1, March.

PCA Publications (1994). Factors Influencing Flow and Strength of Standard Mortars and Reappraisal of the Pozzolanic Activity Index Test, RP 323. Concrete Technology Today, Vol. 15, No. 2, July.

Pepper, L., and Mather. B. (1959). Effectiveness of Mineral Admixtures in Preventing Excessive Expansion of Concrete Due to Alkali–Aggregate Reaction. Proceedings, ASTM 59, pp. 1178–1202.

Perry, C., Day. R.L., Joshi, R.C., Langan, B.W., and Gillott, J.E. (1987). The Effectiveness of Twelve Canadian Fly Ashes in Suppressing Expansion due to Alkali–Silica Reaction. Proceedings, 7th International Conference on Alkali–Aggregate Reaction, pp. 93–97, Ottawa.

Philles, R.E. (1967). Fly Ash in Mass Concrete. Proceedings, First International Symposium on Fly Ash Utilization, Pittsburgh, U.S. Bureau of Mines, Information Circular 1c 8348, pp. 69–79.

Pierce, J.S. (1982). Use of Fly Ash in Combating Sulphate Attack in Concrete. Proceedings, 6th International Symposium of Fly Ash Utilization, Reno, Nevada, DOE/METC/82-52, pp. 208–231.

Pittman, D.W., and Ragan, S.A. (1987). A Guide for the Design and Construction of Roller Compacted Concrete Pavements Paper, Waterways, Experiment Station, U.S. Army Corps of Engineers, Vicksburg, Mississippi.

Popovics, S. (1986). What Do We Know About the Contribution of Fly Ash to the Strength of Concrete. Proceedings: 2nd International Conference on Fly Ash, Silica Fume, Slag and Natural Pozzolans in Concrete. ACI SP-91, pp. 313–332.

Porter, L.C. (1964). Small Proportions of Pozzolan may Produce Detrimental Reactive Expansion in Mortar. Report No. C-113, USBR, 22 pp.

Powers, T.C. (1945). A Working Hypothesis for Further Studies of Frost Resistance of Concrete, Proceedings, Journal of ACI, Vol. 16, No. 4, pp. 245–272.

Price, G.C. (1961). Investigation of Concrete Materials for the South Saskatchewan River Dam. ASTM Proceedings 61, pp. 1155–1179.

Price, W.H. (1951). Factors Influencing Concrete Strength. Proceedings, Journal of ACI, Vol. 47, No. 31, pp. 417–432.

Prusingski, J.R., Fouad, F.H., and Donovan, M.J. (1993). Plant Performance of High Strength Prestressed Concrete Made with Class C Fly Ash. Proceedings: 10th International Ash Use Symposium, ACAA, EPRI TR-101774, Vol. 2, pp. 41-1 to 41-15, January.

Quo, L.W., Lai, C.I., Liu, C.M., and Tsu, C.C. (1993). Porosity and Chloride Penetration of Cement–Fly Ash Mortars. Proceedings: 10th International Ash Use Symposium, ACAA, EPRI TR-101774, Vol. 2, pp. 42-1 to 42-12, January.

Ramachandran, V.S. (1995). Concrete Admixtures Handbook: Properties, Science, and Technology. Second Edition. Noyes Publications, Park Ridge, New Jersey. 1153 pp.

Ramachandran, V.S., Cheung, M.S., and Hachem, H.M. (1990). Low Temperature effects of the Microstructure of Cement Paste Exposed to Sea Water. ACI Materials Journal, Technical Paper, Title No. 87-M36, pp. 340–347, August.

Ramachandran, V.S., Ramakrishnan, V., and Johnston, D. (1992). The Role of High Volume Fly Ash in Controlling Alkali–Aggregate Reactivity. ACI SP-132, Vol. 1, pp. 591–614.

Ramakrishnan, V., Coyle, W.V., Brown, J., Tluskus, A., and Benkataramanyam, P. (1981). Performance Characteristics of Concrete Containing Fly Ash. Proceedings, Symposium on Fly Ash Incorporation in Hydrated Cement Systems. Edited by S. Diamond. Materials Research Society, Boston, pp. 233–243.

Ramezanianpour, A.A., and Malhotra, V.M. (1994). The Effect of Curing on the Compressive Strength, Resistance to Chloride-Ion Penetration and Porosity of Concretes Incorporating Slag or Fly Ash or Silica Fume. Proceedings, 3rd CAN-MET/ACI International Conference on Durability of Concrete, Nice, France, pp. 739–752.

Rasheeduzzafar, Dakhil, and Mukaramm (1987). Influence of the Cement Composition and Content on the Corrosion Behaviour of Reinforcing Steel in Concrete. ACI SP-100, pp. 1477–1502.

Ravina, D. (1981). Efficient Utilization of Coarse and Fine Fly Ash in Precast Concrete by Incorporating Thermal Curing. ACI Journal, Vol. 78, No. 3, pp. 194–200.

Rilem, Technical Committee 12-CRC (1974). Corrosion of Reinforcement and Prestressing Tendons: A State of the Art Report. Materials and Structures, Vol. 9, pp. 187–206.

River, V.R. (1986). Effect of Temperature on the Properties of Mortars and Superplasticized Concrete Containing Low-Calcium Fly Ash. Proceedings, 2nd International Conference on Fly Ash, Silica Fume, Slag, and Natural Pozzolans in Concrete. ACI SP-91, pp. 219–230.

Rodway, L.E., and Fedirko, W.M. (1989). Superplasticized High Volume Fly Ash Structural Concrete. ACI SP-114, Vol. 1, pp. 98–112, June.

Rodway, L.E., and Fedirko, W.M. (1992). Setting Times and Compressive Strength of Superplasticized High Volume Fly Ash Structural Concrete. Fourth CANMET/ACI International Conference on Fly Ash, Silica Fume, Slag and Natural Pozzolans in Concrete. ACI SP-132, Vol. 3, pp. 257–266.

Rosner, J.C. (1976). Let's Design Fly Ash Concretes. Not Compare Them Proceedings, 4th International Ash Utilization Symposium, St. Louis. ERDA, MERC/SP-76/4, pp. 560–572.

Roy, N.K., Murtha, M.J., and Burnet, G. (1978). Recovery of Iron Oxide from Power Plant Fly Ash by Magnetic Separation, IS-M-153.

Ryan, J.P. (1951). Relationship of Fly Ash and Corrosion. ACI Journal, Vol. 47, pp. 481–484.

Ryan, W.G.J., and Munn, R.L. (1979). Some Recent Experiences in Australia With Superplasticizing Admixtures. Proceedings, International Symposium on Superplasticizers in Concrete. ACI SP-62, pp. 123–136.

Sadgrove, B.M. (1974). Freezing of Concrete at Early Age. Technical Report No. 42.503, Cement and Concrete Association, London.

Samarin, A., and Ryan, W.G.J. (1975). Experience in Use of Admixtures in Concrete Containing Cement and Fly Ash. Workshop on the Use of Chemical Admixtures in Concrete.

University of New South Wales, Sydney, Australia, pp. 91–112, December.

Sanjuan, M.A., and Andrade, C. (1994). Effect of Alkali–Silica Reaction on the Corrosion of Reinforcement. 3rd CAN-MET/ACI International Conference on Durability of Concrete, Nice, France, pp. 613–620.

Sarkar, S.L., et al. (1993). Mineral Admixtures in Cement and Concrete, Vol. 4. ABI Books Pvt. Ltd., New Delhi, 565 pp.

Saskatchewan Power Corporation, Canada (1979). Study of Potential Uses of Ash from Fossil Fired Generating Stations. R&D No. 4469-1979 (Ash Utilization) DSS File No. 0455JKE 204-8-1252, 258 pp.

Saucier, K.L. (1980). High Strength Concrete Past, Present, Future. Concrete International, Vol. 2, pp. 46–50, June.

Saucier, K.L. (1982). Use of Fly Ash in No Slump Roller Compacted Concrete. Proceedings, 6th International Symposium on Fly Ash Utilization, Reno, Nevada, pp. 282–293, March.

Schiepl, P., and Hardtle, R. (1994) Relationship Between Durability and Pore Structure Properties of Concretes Containing Fly Ash. K.P. Mehta Symposium on Durability of Concrete. Edited by I.H. Khayad and P.C. Aitcin. Nice, France, pp. 99–118.

Schiepl, P., and Raupach, M. (1992). Influence of the Type of Cement on the Corrosion Behaviour of Steel in Concrete. 9th International Congress on the Chemistry of Cement, New Delhi, Vol. V, pp. 296–301.

Schnieder, U. Verhalten (1982). Von Beton Bei Hogen Temperaturen Dtsch. Ausschuss Stahlb, 337 pp.

Sen, A.K., and De, A.K. (1987). Adsorption of Mercury (II) by Coal Fly Ash. Water Research, Vol. 21, pp. 885–888, August.

Short, N.R., and Page, C.L. (1982). The Diffusion of Chloride Ions through Portland and Blended Cement Pastes, Silicates Industrials 10, pp. 237–240.

Sivasundaram, V., Carette, G.G., and Malhotra, V.M. (1989a). Long Term Strength Development of High Volume Fly Ash Concrete. MSL Division Report, MSL 89-S3, Energy, Mines and Resources, Canada, May.

Sivasundaram, V., Carette, G.G., and Malhotra, V.M. (1989b). Properties of Concrete Incorporating Low Quantity of Cement and High Volumes of Low-Calcium Fly Ash, Proceedings of the 3rd International Conference on Fly Ash, Silica Fume, Slag and Natural Pozzolans in Concrete, ACI SP-114, pp. 45–71.

Sivasundaram, V., Carette, G.G., and Malhotra, V.M. (1990). Selected Properties of High Volume Fly Ash Concretes. ACI Concrete International, October, pp. 47–50.

Smith, Charles L. (1993a). Commercial Aggregate Production from Fly Ash and FGD Waste. Proceedings, 10th International Ash Use Symposium, ACAA, EPRI TR-101774, Vol. 1, pp. 10-1 to 10-13.

Smith, Charles L. (1993b). 15 Million Tons of Fly Ash Yearly in FGD Sludge Fixation. Proceedings, 10th International Ash Use Symposium, ACAA, EPRI TR-101774, Vol. 1, pp. 2-1 to 2-10.

Smith, I.A. (1967). The Design of Fly Ash Concretes. Proceedings Institution of Civil Engineers, London, Vol. 36, pp. 369–790.

Smith, R. (1993). Fly Ash and the Alkali–Aggregate Reaction, What's New? Proceedings, 10th International Ash Use Symposium, ACAA, EPRI TR-101774, Vol. 2, pp. 47-1 to 47-17.

Smith, R.L., and Raba, C.F. Jr. (1986). Recent Developments in the Use of Fly Ash to Reduce Alkali–Aggregate Reaction, Alkalies in Concrete. Edited by V.H. Dodson. ASTM STP 930, Philadelphia, pp. 58–68.

Sourshia, P., and Alhozaimy, A. (1993). Correlation Between Strength and Permeability of Fly Ash Concrete, Proceedings,

10th International Ash Use Symposium, ACAA, EPRI TR 101774, Vol. 2, pp. 38-1 to 38-13, January.

Spinkel, M.M., and Lee. B. (1987). The Chase Air Indicator VHTRC 81-R37. Virginia Highway and Transportation Research Council, Charlottesville, Virginia.

Sprung, S., and Adabia, M. (1970). The Effect of Admixtures on Alkali–Aggregate Reaction in Concrete. Proceedings, 3rd International Conference on Effects of Alkalies on the Properties of Concrete, London. Edited by A.B. Poole, pp. 125–137.

Stanton, T.E. (1942). Expansion of Concrete Through Reaction Between Cement and Aggregate. Transactions ASCE Part 2, pp. 68–85.

Strubble, L., and Diamond, S. (1986). Influence of Cement Alkali Distributions on Expansion due to Alkali–Silica Reaction, Alkalies in Concrete. Edited by V.H. Dodson. ASTM STP 930, Philadelphia, pp. 31–45.

Sturrup, V.R., and Clendenning, T.G. (1969). The Evaluation of Concrete by Outdoor Exposure. Highway Research Board, Record, HRR268, pp. 48–61.

Sturrup, V.R., Hooton, R.D., and Clendenning, T.G. (1983). Durability of Fly Ash Concrete. ACI SP-79, pp. 71–86.

Swamy, R.N. (1986). Cement Replacement Materials. Surrey University Press. 259 pp.

Swamy, R.N., and Mahmud, H.B. (1986). Mix Proportions and Strength Characteristics of Concrete Containing 50 Percent Low-Calcium Fly Ash. Proceedings, 2nd CANMET/ACI International Conference on Fly Ash, Silica, Fume, Slag and Natural Pozzolans in Concrete. ACI SP-91, Vol. 1, pp. 413–432.

Swamy, R.N., Ali, S.A.R., and Theodorakopoulos, D.D. (1983). Early Strength Fly Ash Concrete for Structural Applications. ACI Journal, Vol. 80, No. 5, pp. 414–423.

Swenson, E.G., and Gillott, J.E. (1960). Characteristics of Kingston Carbonate Rock Reaction. Highway Research Board (HRB) Bulletin No. 275, pp. 18–31. HRB, NRC, Washington, DC.

Synder, M.J. (1962). A Critical Review of the Technical Information on the Utilization of Fly Ash. Edison Electric Institute Report, pp. 62–902.

Talero, R. (1993). Sulphitic Characterization of a Pozzolanic Addition. Accelerated Method of the Test to Determine It. (Lechatelier-Ansette and ASTM C452 Methods). Proceedings, 10th International Ash Use Symposium, ACAA, EPRI TR0101774, Vol. 2, pp. 52-1 to 52-10, January.

Tarun, R.N., and Ramma, B.W. (1989). High Strength Concrete Containing Large Quantities of Fly Ash. ACI Materials Journal, pp. 111–116, March–April.

Tattersall, G.H. (1983). The Workability of Fresh Concrete. View Point Publication.

Tattersall, G.H., and Benfill, P.F.G. (1983). The Rheology of Fresh Concrete. Pitman, London.

Tayabji, S.D., and Okamoto, P.A. (1987). Engineering Properties of Roller Compacted Concrete, Annual Meeting of the Transportation Research Board, Washington, DC, January.

Tenney, M.W., and Echelberger, W.F. Jr. (1970). Fly Ash Utilization in the Treatment of Polluted Water. Proceedings: 2nd Ash Utilization Symposium, Pittsburgh, PA, pp. 237–261, March.

Thomas, M.D.A., Gripwell, J.B., and Owens, Philip L. Fly Ash Concrete — An Overview of 30 Years Experience in the United Kingdom. Proceedings, 10th International Ash Use Symposium American Coal Ash Association, ACAA, EPRI TR-101774, Vol. 1, pp. 37-1 to 37-15.

Tikalsky, P.J., Freeman, R.B., and Carrasquillo, R.L. (1990). Recommendations for the Use of Fly Ash in Sulphate Resistant Concrete, CANMET International Workshop on Fly Ash in Concrete, Calgary, Alberta, Canada.

Timms, A.G., and Grieb, W.E. (1956). Use of Fly Ash in Concrete. Proceedings, Vol. 56, ASTM, Washington, DC, pp. 1139–1160.

Torri, K., and Kawamura, M. (1992). Pore Structure and Chloride Permeability of Concretes Containing Fly Ash, Blast Furnace Slag and Silica Fume, Proceedings: 4th International Conference on Fly Ash, Silica, Fume, Slag and Natural Pozzolans in Concrete. ACI SP-132, pp. 135–150.

Transportation Research Board (1986). Use of Fly Ash in Concrete. National Co-Operative Highway Research Program Synthesis of Highway Practice, 127, Washington, DC, 66 pp.

Tsukayama, R., Tagataki, S., and Abe, H. (1980). Long Term Experiments on the Naturalization of Concrete Mixed with Fly Ash and Corrosion of Reinforcement. Proceedings, 7th International Congress on the Chemistry of Cement, VII-ISCC, Paris.

Tynes, W.O. (1982). Fly Ash and Water Reducing Admixtures for Articulated Concrete Mattress. Miscellaneous Papers 6-473, Army Corps of Engineers, WES, Vicksburg, Mississippi.

Tysl, S.R., Farzam, H., and Factor, D. (1994). A Chemical Admixture for Control of Alkali–Silica Reaction. Proceedings, 3rd CANMET/ACI International Conference on Durability of Concrete, Nice, France, pp. 585–598.

Uchikawa, H., Uchida, S., and Ogawa, K. (1982). Influence of the Properties of Fly Ash on the Fluidity and Structure of Fly Ash Cement Paste. Proceedings, International Symposium on the Use of PFA in Concrete. University of Leeds, England, pp. 83–95, April.

Uchikawa, H., Uchida, S., and Ogawa, K. (1983). Influence of Superplasticizer on the Hydration of Fly Ash Cement. Silicate Industrials, Vol. 4, pp. 99–106.

Ukite, K., Ishii, M., Yamamoto, K., Azumak, K., and Kolino, K. (1992). Properties of High Strength Concrete Using Classified Fly Ash. Proceedings: 4th International Conference on Fly Ash, Silica Fume, Slag and Natural Pozzolans in Concrete. ACI SP-132, pp. 37–52.

U.S. Department of Energy (1985). Proceedings of the Seventh International Ash Utilization Symposium and Exposition. DOE/METC-85-6018. National Technical Information Services, U.S. Department of Commerce, Springfield, Virginia, May.

Vandewalle, L., and Mortelmans, F. (1992). The Effect of Curing on the Strength Development of Mortar Containing High Volumes of Fly Ash. Proceedings: 4th International Conference on Fly Ash, Silica Fume, Slag and Natural Pozzolans in Concrete. ACI SP-132, pp. 53–62.

Virtamen, J. (1983). Freeze Thaw Resistance of Concrete Containing Blast Furnace Slag, Fly Ash or Condensed Silica Fume, Proceedings: 1st International Conference on the Use of Fly Ash, Silica Fume, Slag and Other Mineral By-Products. ACI SP-79, pp. 923–942.

Vivaraghavan, T., and Rao, G.A.K. (1991). Removal of Cadmium and Chromium from Waste Water Using Fly Ash. 45th Purdue Industrial Waste Conference Proceedings, Lewis Publishers Inc., Chelsea, M148118.

Vivian, H.E. (1976). Alkalies in Cement and Concrete, CSIRO Division of Building Research Highest, Victoria, Australia. Proceedings: 3rd International Conference on Effects of Alkalies on the Properties of Concrete, pp. 9–23, September.

Volkwein, A. (1994). Possible Penetration of Gases, Fluids and Salts Due to the Internal Shrinkage. Proceedings, 3rd CAN-

MET/ACI International Conference on Durability of Concrete, Nice, France, pp. 599–612.

Von Fay, K.F., and Pierce, J.S. (1989). Sulphate Resistance of Concretes with Various Fly Ashes, ASTM Standardization News, pp. 32–37, December.

Von Fay, K.F., Kepler, W.F., and Drahushak-Crow, R. (1993). Freeze–Thaw Durability of Concretes with Various Fly Ashes, Proceedings, 10th International Ash Use Symposium. ACAA, EPRI TR-101774, Vol. 2, pp. 40-1 to 40-14, January.

Wackerla, E. (1991). Premature Freezing of Fly Ash Concrete, MEng Thesis, University of Calgary, Calgary, Alberta, Canada, 115 pp., September.

Wei, Libua, Naik, Tarum, R., and Golden, Dean M. (1993). A Study on the Properties of Coal Ash Masonry Blocks. Proceedings, 10th International Ash Use Symposium ACAA, EPRI TR-101774, Vol. 2, pp. 78-1 to 78-10, January.

Welsh, G.B., and Burton, J.R. (1958). Sydney Fly Ash in Concrete. Commonwealth Engineer, Australia, pp. 62–67, January.

Weng, C.H. (1990). Removal of Heavy Metals by Fly Ash. MSc Thesis, University of Delaware.

Wentz, C.A., Moretti, C.J., Henke, K.R., Manz, O.E., and Wilken, K.U. (1988). Use of Fly Ash as a Waste Minimization Strategy. Environmental Progress, Vol. 7, No. 3, pp. 198–206, August.

White, T.D. (1986). Mix Design, Thickness Design and Construction of Roller Compacted Concrete Pavement. Transportation Research Record 1062, TRB.

Whiting, D. (1990). Strength and Durability of Lean Concrete Containing Fly Ash. Research and Development Bulletin RD 099T, PCA.

Whiting, D., and Stark D. (1983). Control of Air Content in Concrete. National Co-Operative Highway Research Program (NCHRP) Report 258, Transportation Research Board (TRB).

Wolsiefer, J. (1982). Ultra High Strength Field Placeable Concrete in the Range 10,000 to 18,000 psi (69–124 MPa). Paper presented at ACI Convention, Atlanta, Georgia, pp. 19–23, January.

Xu, A., Chandra, S., and Rhode, M. (1994). Influence of Alkali on Carbonation of Concrete. Proceedings: 3rd CANMET/ACI International Conference of Durability of Concrete, Nice, France, pp. 173–182.

Yuan, R.L., and Cook, J.E. (1983). Study of a Class C Fly Ash in Concrete. Proceedings, 1st International Conference on the Use of Fly Ash, Silica Fume, Slag, and Other Mineral By-Products in Concrete. ACI SP-79, pp. 307–319.

INDEX

A

Abrasion/erosion resistance of concrete, 152–153

Accelerators as admixtures in fly ash concrete, 174–175

Acid mine drainage neutralization, 208–209

Acid rain, 214

Acid solutions, 119

Addition method for concrete mix proportioning, 58–59

Admixtures in fly ash concrete, 173–190

Adsorption by fly ash, 206–207

Aerated concrete, 191

Air content and air entrainment, 81

Air content test, 61

Air entraining agents, 40–41, 73, 79–83, 185–187

Air entrainment, 39–41, 63–65, 68, 79–83, 95
  and freeze–thaw durability, 114, 115
  of roller compacted concrete, 157, 158

Alberta fly ash
  and air entrainment, 80
  and alkali–silica reactions, 138–139
  and cement setting, 77–79
  and freeze–thaw durability, 115–116
  and strength development, 93–96
  and sulphate resistance, 122–123
  in structural concrete, 161, 166–167

Alkali–carbonate reaction, 137

Alkalis
  and cement durability, 132–134
  in fly ash, 42

Alkali–silica reaction, 130–142

Alumina, 10, 11
  and sulphate resistance, 125–126

Alumina oxide, 34–36, 38

Alumino-silicates, 15, 24, 54

American Coal Ash Association (ACAA), 6

Anhydrite, 45–46

Anthracite coal, 16

X

Y